Molecular Medical Science Series

Molecular Genetics of Human Inherited Disease

Edited by
DUNCAN J. SHAW
The University of Aberdeen,
Scotland, UK

JOHN WILEY & SONS
Chichester · New York · Brisbane · Toronto · Singapore

Other Wiley Editorial Offices

John Wiley & Sons, Inc., 605 Third Avenue,
New York, NY 10158-0012, USA

Jacaranda Wiley Ltd, 33 Park Road, Milton,
Queensland 4064, Australia

John Wiley & Sons (Canada) Ltd, 22 Worcester Road,
Rexdale, Ontario M9W 1L1, Canada

John Wiley & Sons (SEA) Pte Ltd, 37 Jalan Pemimpin £05-04,
Block B, Union Industrial Building, Singapore 2057

Library of Congress Cataloging-in-Publication Data

Molecular genetics of human inherited disease / edited by Duncan J.
   Shaw.
       p.   cm.—(Molecular medical science series)
       Includes bibliographical references and index.
       ISBN 0 471 93459 3
       1. Genetic disorders.   I. Shaw, Duncan J.   II. Series.
       [DNLM: 1. Herediatary Diseases—genetics.   2. Metabolism, Inborn
   Errors—genetics.   QZ 50 M7186 1995]
   RB155.5.M65   1995
   616'.042—dc20
   DNLM/DLC
   for Library of Congress   95–10485
                                                             CIP

British Library Cataloguing in Publication Data

A catalogue record for this book is available from the British Library

ISBN 0 471 93459 3

Typeset in 10/12 Palatino by Vision Typesetting, Manchester
Printed and bound in Great Britain by Biddles Ltd, Guildford, Surrey

This book is printed on acid-free paper responsibly manufactured from
sustainable forestation, for which at least two trees are planted for each
one used for paper production.

*Molecular Genetics
of Human Inherited Disease*

Molecular Medical Science Series

Series Editors

Keith James, University of Edinburgh Medical School, UK
Alan Morris, University of Warwick, UK

Forthcoming Titles in the Series

Other Titles in the Series

Contents

Contributors

Duncan J. Shaw
Department of Molecular and Cell Biology, University of Aberdeen, Marischal College, Aberdeen AB9 1AS, UK

Helen G. Harley
University of Wales College of Medicine, Institute of Medical Genetics, Heath Park, Cardiff CF4 4XN, UK

Angus Clarke
University of Wales College of Medicine, Institute of Medical Genetics, Heath Park, Cardiff CF4 4XN, UK

J. David Brook
Department of Genetics, University of Nottingham, University Park, Nottingham NH7 2RD, UK

Nick Thomas
University of Wales College of Medicine, Institute of Medical Genetics, Heath Park, Cardiff CF4 4XN, UK

Michael J. Owen
University of Wales College of Medicine, Institute of Medical Genetics, Heath Park, Cardiff CF4 4XN, UK

Philip J. Asherson
University of Wales College of Medicine, Institute of Medical Genetics, Heath Park, Cardiff CF4 4XN, UK

Jeremy Cheadle
University of Wales College of Medicine, Institute of Medical Genetics, Heath Park, Cardiff CF4 4XN, UK

David L. Nelson
Institute of Molecular Genetics, Baylor College of Medicine, Houston, Texas 77030, USA

Jean Weissenbach
Unité de Génétique Moléculaire Humaine, CNRS URA 1445, Institut Pasteur, 25 rue du Dr Roux, 75724 Paris CEDEX 15, France

David Viskochil
Division of Medical Genetics, Department of Pediatrics, University of Utah School of Medicine, Salt Lake City, Utah 84112, USA

1 Introduction

DUNCAN J. SHAW

Medical genetics has moved rapidly from being a largely descriptive science, to one in which molecular analysis plays a key role, and in which therapeutic intervention based on the molecular data is now being attempted. There are two kinds of expertise that have made this progress possible. The first is that of the clinical geneticist, who examines patients and their families to establish whether the condition is genetic, how it is inherited, and what is its relationship to other known conditions. The second is that of the molecular biologist, who has applied an arsenal of techniques originally developed for the study of basic biological processes in lower organisms, to human inherited disease. Sometimes, the two kinds of expertise coexist in the same person; usually collaboration is necessary.

The idea of using genetic markers to map the genes causing human inherited disease is not particularly new, but because the original studies were limited to the use of blood group and protein polymorphisms, which are few in number relative to the size of the human genome, little progress could be made. In 1975, Ed Southern (then in Edinburgh) invented a technique that enabled specific sequences in complex genomes to be visualized using a radioactive complementary DNA fragment as a probe (Southern blotting). Together with developments in molecular cloning, by which large quantities of any fragment of DNA could be obtained in a pure form, this made it possible to study the overall structure of individual genes.

Blood group and protein polymorphisms are caused by genetic variation in the parts of genes coding for proteins. However, the vast majority of the human genome is non-coding. The key to unlocking the real power of genetic linkage for finding disease genes came with the realization, first articulated by David Botstein and colleagues in 1980, that the new methods could be used to study genetic variation in all types of DNA sequence. Any change in sequence that affected the cutting site for a restriction enzyme, or an insertion or deletion of DNA, would cause a change in the length of the associated DNA fragments. This could be detected by Southern blot analysis using a cloned piece of human DNA (coding or non-coding) as a probe. These variants were termed restriction fragment length polymorphisms (RFLPs). The large number of restriction enzymes that were becoming available

Molecular Genetics of Human Inherited Disease. Edited by D.J. Shaw
Published 1995 by John Wiley & Sons Ltd

increased the chances of finding RFLPs, and subsequent studies showed that the number of RFLPs was likely to be sufficient that at least one would be within mapping range of any unknown gene. Thus, it should be possible to locate a disease gene by linkage analysis using DNA from affected families. Once an RFLP that showed statistically significant linkage to the disease was found, the location of the gene would be revealed by mapping the DNA probe that detects the RFLP to a particular region of the human genome. DNA sequences are relatively easy to map, the most direct method being *in situ* hybridization (where a fluorescently labelled probe is hybridized to a metaphase spread of human chromosomes, and the resulting signal is detected by light microscopy).

In order to apply this procedure to human genetic disease, the skills of the clinical geneticist were required. Unlike those of mice and fruit flies, human families have only small numbers of offspring, and are never contrived for the convenience of the geneticist. Individual human pedigrees seldom give conclusive linkage information, but have to be combined with others to 'bump up' the statistics. It is essential therefore that all the families should represent the same clinical condition. The clinical geneticist will apply strict diagnostic criteria and look for any signs of phenotypic heterogeneity, which might suggest that the underlying genetic defect is in different genes in different families. Attempts to combine genetically heterogeneous families in linkage analysis will usually lead to loss of information, unless it can be established which families are of which type and analyse them separately.

During the early to mid-1980s much effort went into locating the genes for the more common inherited disorders, such as Duchenne muscular dystrophy (DMD), Huntington's disease (HD) and cystic fibrosis. These and other important genes are the subjects of the following chapters. Because new RFLPs were constantly being found, the disease genes could be mapped within a reasonable period of time. Linked RFLPs also have a second important application—they may be used for pre-symptomatic and pre-natal diagnosis of the inherited disease. This procedure, which should always be done within the context of a proper programme of family counselling and support, is now in routine clinical use.

However, localizing the gene is only the first step. The real goal is to isolate the gene, find out what it codes for, and establish how it is mutated in patients. This should lead to an understanding of the molecular pathology. The methods of molecular biology, which had mostly been developed for work on organisms with relatively small genomes, needed to be extended to cope with the megabases of DNA that might lie between the RFLPs linked to a disease gene. In this phase of the work, the development of new technology proceeded along with its application to finding the genes. Crucial methods included pulsed-field gel electrophoresis (PFGE), yeast artificial chromosome (YAC) cloning, cosmid contig analysis, radiation hybrids and various methods for identifying coding sequences, which are usually embedded within large stretches of non-coding DNA. The polymerase chain reaction (PCR) has also made a great impact in many areas of molecular biology. Human molecular genetics is no exception, and PCR-based methods for the analysis of gene sequences and mutations are in widespread use. A new class of polymorphic DNA sequence, easily and quickly analysed by PCR, has largely replaced RFLPs as markers for linkage analysis. These are tandemly repeated copies of simple sequences such as $(CA)_n$ and $(CAG)_n$, which vary in copy number between individuals. These markers are also highly abundant and widely dispersed throughout the genome, and have been used to construct the first complete linkage maps of the human genome (Gyapay *et al.*, 1994).

Many important human disease genes have now been isolated, as will be seen from the following chapters. Some of these genes are very large, such as the DMD gene; others, like the myotonic dystrophy (DM) gene, are quite small. In many cases the function of the gene could be guessed from examination of the sequence. For example, the DM gene appears to code for a protein kinase, the cystic fibrosis (CF) protein sequence looked like that of a transmembrane protein, and the neurofibromatosis type 1 gene product is similar to a type of protein involved in intracellular signal transduction. The function of other gene products, such as those for DMD and HD, were not revealed immediately. Subsequent studies of the DMD protein suggested that it is associated with the cytoskeleton. At the time of writing it was too early to say what the function of the HD protein might be.

These genes also show considerable differences in the ways in which they are mutated in affected individuals. The DMD gene is usually disrupted by substantial deletions; the commonest CF mutation is a 3 bp deletion, and most of the remainder are single base changes; and in most cases, it is still not clear how the NF1 gene is mutated. But by far the most surprising result to have emerged from work on these genes is the mechanism of mutation by expansion of tandemly repeated trinucleotides. In the fragile X syndrome the repeat is $(CGG)_n$, located in the 5' untranslated region of the FMR1 gene; in myotonic dystrophy, it is $(CTG)_n$, located in the 3' untranslated region of the DM protein kinase gene; and in HD there is a $(CAG)_n$ repeat in the coding sequence of the gene which is believed to be translated as poly-glutamine. In all cases the difference between a healthy individual and one with the disease is the number of repeats in the sequence. From a few up to around 40 or 50, there is no problem; larger repeats bring with them the disease phenotype, and the larger the repeat, the more severe the symptoms. The most remarkable aspect of these mutations is that they are often genetically unstable, so that when passed on from an affected parent to his or her offspring, further expansion takes place. This instability, which can be dramatic in the case of fragile X and myotonic dystrophy, is now known to be the basis of the phenomenon of 'anticipation'. The latter is the tendency of affected parents to have offspring that are more severely affected, and with earlier onset of the disease, than themselves. Several other examples of inherited neurological disease caused by trinucleotide repeat expansion have since been discovered.

The approaches described in this book, which have enabled disease genes to be isolated without prior knowledge of their structure or function, have become known as 'positional cloning' because the major criterion used to find the gene is its position in the genetic map. The strategy has now been successful in finding the genes for the more common conditions showing simple Mendelian inheritance. However, there are far larger numbers of people that suffer from conditions in which inheritance plays an important part, but is not the sole determinant of disease status. Cancer, heart disease and many psychiatric conditions come into this category. Many medical geneticists believe that the major challenge of the coming years will be to apply positional cloning to finding the genetic factors at work in multifactorial disease. The way in which this is being attempted is the subject of our chapter on psychiatric disorders. The key factors here are the development of a detailed linkage map of the human genome, based on hundreds of DNA markers, and sophisticated statistical methods for disentangling the genetic factors from the background of environmental influences. Genetics is now poised to make a major contribution to our understanding of abnormal psychology.

As well as its success in uncovering what is going wrong in inherited disease, molecular genetics is now beginning to have an impact on treatment. Of the diseases covered in this book, cystic fibrosis is the one in which gene therapy is most advanced, and clinical trials are now under way. Politicians and health service managers notwithstanding, it seems that we can look forward to a new era in molecular medicine, where intervention at the level of the gene will provide new opportunities for conquering some of the most intractable conditions known to medicine.

REFERENCES

Southern EM (1975) Detection of specific sequences among DNA fragments separated by gel electrophoresis. *J. Mol. Biol.*, **98**, 503–517.
Botstein D, White RL, Skolnick M and Davis RW (1980) Construction of a genetic linkage map in man using restriction fragment length polymorphisms. *Am. J. Hum. Genet.*, **32**, 314–331.
Gyapay G, Morissette J, Vignal A *et al.* (1994) The 1993–94 Genethon human genetic linkage map. *Nature Genet.*, **7**, 246–339.

2 Xp21 (Pseudohypertrophic) Muscular Dystrophy: Duchenne and Becker Muscular Dystrophies

ANGUS CLARKE AND NICK THOMAS

CLINICAL BACKGROUND

The disorder that we now recognize as Duchenne muscular dystrophy (DMD) was identified during the nineteenth century as a distinct cause of progressive muscular weakness in boys. The studies of Edward Meryon, Duchenne de Boulogne, William Gowers, Wilhelm Erb and others (Further Reading: Emery, 1993) built up a picture of the condition as affecting muscle tissue directly, rather than as a secondary (neurogenic) effect of a disturbance of the nervous system. The progressive replacement of muscle fibres by connective tissue elements (fat and fibrous tissue) was demonstrated, explaining the progressive weakness. The sex-linked inheritance of DMD through carrier women to affect their sons and their daughters' sons was also recognized. Now, a century or so later, a much clearer picture of the condition has emerged from a combination of clinical and molecular genetic studies.

Boys with DMD usually appear healthy in infancy, although some may be floppy or fail to thrive, but they then show signs of delayed motor development. The gait of many affected boys is noticed by the family to be unusual from about 2 years, although the diagnosis may not be made until several more years have passed unless the relevant health professionals are alert to the diagnosis. There is usually generalized developmental delay, particularly in the locomotor and language areas. This neuropsychological impairment is non-progressive, and the boys' mean IQ scores are in the range 80–89. The verbal IQ score may be 10 points or so lower than the non-verbal score. The boys' locomotor performance falls progressively further behind, whereas they make average progress in other areas.

Some muscle groups, particularly in the calf, may appear bulkier than normal (pseudo-hypertrophy) before any weakness is apparent; other muscles, such as the quadriceps in the thigh, may show wasting. The distribution of muscle weakness and wasting is predominantly proximal,

Molecular Genetics of Human Inherited Disease. Edited by D.J. Shaw
Published 1995 by John Wiley & Sons Ltd

affecting the muscles of the pelvic and shoulder girdles earliest and most severely. A cardiomyopathy may develop, and there are often changes apparent on the ECG; indeed, some cases of X-linked dilated cardiomyopathy are associated with deletions of the DMD gene promoter but do not exhibit generalized muscle weakness. Smooth muscle dysfunction may also be a feature, resulting in gastric hypomotility. The skeletal muscle weakness leads to progressive difficulty walking up a slope or climbing stairs, rising from a chair or running. Most boys become confined to a wheelchair between the ages of 8 and 12 years. Tendon contractures (resulting in joint deformities) and scoliosis (spinal curvature restricting proper inflation of the lungs with breathing) may be major problems. Once unable to walk, boys usually experience a long period of gradual decline in physical strength, culminating in death from chest infection or, less frequently, from the cardiomyopathy; however, survival has been improving and is now often 15–25 years of age, and occasionally even longer.

Becker muscular dystrophy (BMD), now known to be caused by mutations at the same gene locus, was initially distinguished from DMD by Becker in 1955 on account of its milder clinical course. Affected males are usually still ambulent at 16 years, and may never require a wheelchair. There is a great range of severity covered by this diagnosis, so that some cases appear to have a myopathy confined to the quadriceps or a non-progressive myalgia with cramps but without weakness. Whether these conditions can all be called BMD has become a purely semantic question, now that the underlying gene defects are known to be allelic. There is also a small group of Xp21 dystrophy patients with features intermediate between DMD and the severe end of the BMD spectrum.

The distribution of muscle involvement in BMD is similar to that in DMD, but the rate of progression is slower. Although the birth incidence of BMD is about one-third that of DMD, the prevalence in the population is similar because survival is so much better in BMD. There is a similar tendency to a discrepancy between the verbal and performance components of the IQ to that found in DMD, and one in 10 boys may be assessed as educationally subnormal.

The clinical judgement that BMD and DMD were distinct disorders was supported by pathological studies that revealed 'neurogenic' changes in BMD, as opposed to the myopathic features in DMD muscle, and by some classical genetic linkage studies suggesting linkage with colour-blindness and glucose 6-phosphate dehydrogenase (G6PD); colour-blindness and G6PD were known to be closely linked to each other and not to DMD. It later became clear that these results were misleading for two reasons: first, some of the families included in the linkage studies were subsequently shown to have Emery–Dreifuss muscular dystrophy—a distinct form of X-linked muscular dystrophy which is now known to map close to the red–green colour vision genes at Xq28; second, some cases of BMD-type severity had previously been diagnosed as having 'X-linked spinal muscular atrophy with calf hypertrophy' on the basis of electrophysiological and histopathological studies, but were found to have molecular deletions in the DMD gene. Such families had been excluded from some early studies of BMD.

In the absence of a rational, curative therapy, the medical management of boys with DMD has been largely supportive. Physiotherapy helps to prevent tendon shortening and consequent joint contractures. Dietetic advice helps to prevent obesity. Occupational therapists help a boy to achieve as much as possible given the particular level of his physical abilities. Orthopaedic surgeons can operate to correct or to prevent the progression of a scoliosis, and to lengthen tendons, such as the Achilles tendon, if they become shortened. Respiratory care physicians and anaesthetists have set up programmes for the respiratory care of patients with neuromuscular disorders, including DMD; for selected patients with severe symptoms of respiratory impairment, nocturnal ventilation at home can improve the quality of life considerably.

There are three additional therapies that have been developed, and which some families and professionals believe to be efficacious. The early tendon transfer surgery of Rideau, carried out at 4–6 years, is claimed to prolong ambulation in DMD boys. The administration of prednisone is thought by some to produce an improvement in muscle strength at least in the medium term, although side-effects are commonly encountered and there must be concern about using as a long-term treatment for muscle disease a drug that frequently causes muscle wasting, particularly since these studies have employed historical rather than contemporaneous controls and their methodology has therefore not been randomized or double-blind. Finally, one boy with both DMD and growth hormone deficiency suffered a relatively slow deterioration; this raised the question of whether height might be related to the rate of progression of DMD. However, therapeutic trials with a growth hormone inhibitor failed to demonstrate any consistent effect. No rational, curative treatment known to be safe and effective is available at present.

TOWARDS THE DUCHENNE GENE

EARLY LINKAGE

The first direct application of the recombinant molecular genetic techniques to DMD was the identification of a cloned anonymous DNA sequence from the short arm of the X chromosome. This marker, lambda RC8 (DXS9), was found to be linked to the DMD locus; this was consistent with two reports of girls clinically affected by DMD, who had X;autosome translocations in both of which the X breakpoint was in the same region of the X chromosome short arm.

Another marker soon became available, L1.28 (DXS7). This and DXS9 were

located 15–20 cM from the DMD gene, and on opposite sides (flanking) (Fig.
2.1a). At the same time, BMD was also found to be linked to DXS7 and DXS9,
so it appeared likely that the DMD and BMD loci would be close, and perhaps
identical. This helped to resolve the clinical definitions of the different
X-linked muscular dystrophies.

Further work identified closer markers, obtained from genomic, X chromosome-specific and cDNA libraries, such as p754 (DXS84) and C7 (DXS28). By
this time, several more cases of DMD/BMD in females with X;autosome
translocations had been reported (Boyd *et al.*, 1986), as well as a cytogenetic
microdeletion of Xp21 found in a male with DMD. All these translocation
cases had X breakpoints at Xp21, but spread over a broad segment of this
band, suggesting that the DMD gene might be very large.

pERT87 AND XJ

Progress then accelerated in the wake of the identification of some atypical
cases of DMD/BMD. Uta Francke reported a boy, 'BB', who suffered from a
cluster of X-linked disorders including DMD and chronic granulomatous
disease (CGD), an immunodeficiency affecting neutrophil killing of bacteria.
He had a cytogenetic deletion of part of band Xp21 (Fig. 2.1b). Lou Kunkel
and his team then set out to isolate the DNA sequences that were deleted in
BB. To achieve this, they fragmented BB's DNA by shearing, which leaves
blunt ends, and mixed a large excess of this with a small quantity of DNA
from a 49,XXXXY lymphoblastoid cell line. The DNA from the 4X-cell line
had been digested with a restriction endonuclease, MboI, that leaves sticky
ends that permit ready cloning and ligation. The DNA in the mixture was de-
natured, and extensively reassociated using a technique of phenol enhancement (PERT). The vast majority of the sequences from the 4X-cell line should
anneal to BB's DNA, because that was present in large excess. However, those
sequences present in the 4X-cell line, but absent from BB's DNA because of
the deletion, could only anneal to their complementary strands with sticky
ends; such sequences were then selected by cloning into a suitable plasmid
vector.

Derivative clones from this deletion-specific library were identified by
somatic cell hybrid techniques as recognizing sequences around Xp21, and by
their failure to recognize BB's DNA. These clones, designated as pERT87
subclones, were then screened against DNA samples from boys with DMD
and their families. Pooled results from many centres (Kunkel *et al.*, 1986)
indicated that 6.5% of DMD/BMD cases had deletions detected by three
subclones of pERT87, and that their associated RFLPs demonstrated recombination between the pERT subclones (DXS164) and the disease phenotype in
about 5% of meioses.

The Xp21 DNA sequences absent from the cell line of BB were subsequently
also employed to identify expressed transcripts from this region specific to

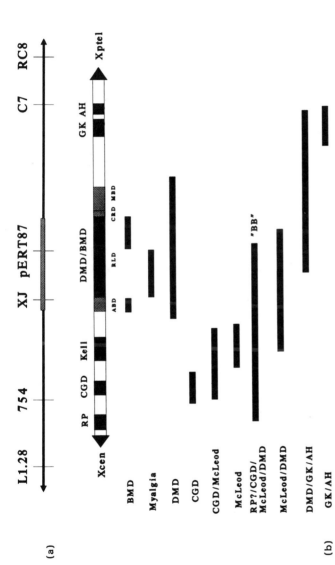

Fig. 2.1. (a) The framework genetic marker map of the Xp12–p22 region identifying the polymorphic marker loci initially used to localize the DMD/BMD gene to this region. (b) A physical map of the Xp21.1–p21.3 region illustrating the numerous genes which flank the dystrophin gene, and the often complex inherited diseases that can result from different DNA deletions within this gene-rich region. The genes are: RP (retinitis pigmentosa), CGD (chronic granulomatous disease), Kell (McLeod syndrome), DMD/BMD (dystrophin), GK (glycerol kinase deficiency) and AH (adrenal hypoplasia). Interstitial deletions which remove part, or all, of one of these gene loci will produce only that particular disease. Those deletions which involve two or more of these genes, however, usually result in a *contiguous gene syndrome*. Thus the original male patient 'BB', with the large X chromosome deletion involving the retinitis pigmentosa, chronic granulomatous disease, Kell and dystrophin genes (RP/CGD/McLeod/DMD), expressed a spectrum of clinical symptoms characteristic of all four diseases.

phagocytic cells, and this greatly facilitated the cloning of the gene for CGD.

While Kunkel and colleagues were studying BB, Ron Worton and his colleagues in Toronto were investigating a mentally retarded teenage girl with an X;21 translocation and a muscular dystrophy resembling DMD/BMD. As originally suggested by Lindenbaum when he first described such a female in 1979, the translocation breakpoint on Xp was likely physically to disrupt the DMD locus. Since the intact X chromosome is preferentially inactivated in such X;autosomal translocations, the woman is left with little or no functioning DMD gene. Ray, Worton and colleagues (Ray et al., 1985) set out to clone the X;21 breakpoint on chromosome 21, which lay within a region of repeated transcription units of ribosomal RNA (rRNA) genes. They separated the two reciprocal derivative chromosomes into a mouse background in somatic hybrid cell lines, so that the rRNA genes elsewhere in the human genome would be removed. They orientated the rRNA genes on the der(X) chromosome, and produced libraries of sequences from this cell line in a variety of vectors. In these, they searched for clones containing sequences from one end of the rRNA gene cluster as well as sequences from the human X chromosome. Unfortunately these initial searches failed. They subsequently produced probes specific for human rDNA, by a comparison of the 28 S rRNA gene sequences from the human and mouse. This enabled them to identify several unique fragments from libraries from the der(X) cell line, which had not appeared in their previous searches. One of these fragments did indeed contain the X;21 breakpoint and was finally isolated by cutting the electrophoresis gel and cloning all DNA fragments of the appropriate size. This clone (XJ1-1, DXS206) also revealed molecular deletions in several males with DMD.

DMD GENE TRANSCRIPTS

The cDNA product of the DMD locus was first identified by Kunkel's group in Boston (Monaco et al., 1986). They searched for sequence conservation between species, especially between humans and mice. Only two subclones of pERT87 (DXS164)—pERT87-25 and pERT87-4— hybridized strongly to all the mammalian genomes tested. Sequence homology between the human and the corresponding mouse sequences was demonstrated. Open reading frames were sought: only one was found in both subclones in both species that also incorporated the appropriate splice site recognition sequences, with the same orientation in both subclones. The DMD gene was shown to be orientated 5' proximally at Xp21.2 in the human.

Human fetal mRNA samples from muscle and other tissues were screened, and an expressed component was recognized by the pERT87-25 subclone in fetal skeletal muscle as well as some tissues containing smooth muscle, but not brain. Confirmation was achieved by constructing a cDNA library from the fetal muscle mRNA sample; this was screened with the pERT87 subclones

25 and 4; pERT87-25 identified several different cDNA clones. These clones, when hybridized back to fetal muscle RNA, identified mRNA of the same size. The total genomic sequences identified by these overlapping cDNA clones covered more than 100 kb. Because of the unusually high intron/exon ratio, the DMD gene was predicted to be 1–2 Mb in length.

Worton's group subsequently demonstrated that the DXS206 locus (recognized by the XJ1-1 clone) also included conserved sequences that are transcribed in adult muscles. Adult muscle cDNA included an mRNA containing exons from DXS206, as well as from the DXS164 locus, and from sequences proximal to both loci (Burghes *et al.*, 1987). The DMD-specific mRNA was initially thought to be approximately 16 kb, but this was later modified to be 14 kb in size.

PHYSICAL MAPPING OF THE DMD GENE REGION

Restriction fragment maps of the entire DMD gene region were constructed using several different enzymes to clarify the overall genomic structure of the locus. These studies revealed further polymorphisms and permitted the accurate identification and sizing of the deletions in affected males. The clarification of exon structure in the 3' region of the gene proved difficult, but has recently been achieved with the selective sequencing of intron/exon boundaries by means of the Vectorette PCR technique applied to regions of the DMD gene inserted into yeast artificial chromosomes (YACs) (Roberts *et al.*, 1992). There are at least 79 exons in the dystrophin gene, and the whole gene encompasses some 2.3 Mb of DNA.

The use of pulsed-field gel electrophoresis (PFGE) permits the resolution of DNA fragments of up to several megabases in size. In this way, using only a restricted range of DNA markers, it is possible to detect structural rearrangements within the DMD gene. Thus deletions of DNA sequences not recognized directly by the marker, but present on a large PFGE restriction fragment that is recognized by the marker, can be detected. If a length of DNA is deleted from the region of the marker (but not including the marker sequence), the marker will then detect a fragment of altered size—a junction fragment. The presence of such junctional fragments can be used to identify female carriers of DMD in families carrying such deletions.

A large-scale physical map of the DMD region has been produced by means of PFGE. This work has been facilitated by the large number of DNA markers in the area, and by the fact that Xp21 is dark-staining on Giemsa and contains correspondingly few ummethylated CpG sites and HTF (HpaII tiny fragment) islands, often found at the 5' end of 'housekeeping' and other genes. Thus, the DMD region produces only a few large fragments when rare-cutting restriction enzymes are used for PFGE analysis—and this makes the construction of a physical map much simpler than if many, smaller fragments were produced.

The physical map of the DMD gene region incorporated many expressed loci, both proximal and distal to the DMD locus, such as the glycerol kinase (GK) and congenital adrenal hypoplasia (CAH) loci. GK deficiency (GKD) (itself usually harmless) was known to cosegregate with CAH in some families, and these two conditions were found in association with DMD in several families; in other families ornithine carbamoyltransferase (OCT) deficiency or retinitis pigmentosa (RP) was present with DMD in the affected male. A cytogenetic deletion has been identified in some of these contiguous gene deletion syndrome cases (Fig. 2.1b). These studies suggested strongly that OCT would lie proximal to DMD, and that CAH and GKD should lie close to each other, on the opposite side of DMD from the loci for CGD and RP. The more recent, detailed molecular studies of contiguous gene deletion cases have indeed confirmed that the loci for CAH and GKD do lie distal to the DMD locus, while CGD and RP lie proximal.

MOLECULAR STUDIES OF MUTATIONS IN THE DMD GENE

DELETIONS

Initial studies had detected fewer DMD gene deletions in males with BMD than with DMD, but this subsequently proved to have been misleading. Davies' group reported a 22% rate of deletions in cases of DMD/BMD detected with the cDNA clone Cf23a; they found that many of the deletions in BMD families started in the same region of the cDNA, 3' to the regions that had previously been studied. Using this marker in addition to the previously described markers pERT87, XJ1-1 and HIP25, deletions could be found in 40% of cases. When the 3' region of DNA in which deletions could be detected was extended by the use of another cDNA clone, 67% of DMD/BMD cases were found to have deletions. Attempts to correlate disease characteristics such as severity, mental retardation or short stature with the site or extent of the deleted region were generally unsuccessful, although a few specific patterns of deletion were found solely, or predominantly, in individuals with either BMD or DMD.

A clustering of deletions was identified in two areas of the DMD gene: at the 5' end and within the central region (Fig. 2.2). There is a specific deletion hot spot in intron 44, a large intron of 146 kb containing the marker P20 and in which 30% of all DMD gene deletions have one breakpoint. A large number of breakpoints in this intron have been localized precisely, to the nearest restriction site, and they have been found to be distributed evenly over the intron. This makes it less likely that a few specific sequences are initiating the rearrangements, and more likely that structural characteristics or a more diffuse tendency to rearrangement account for the high rate of deletions. Runs of AT-rich sequences have been found close to the breakpoints in some

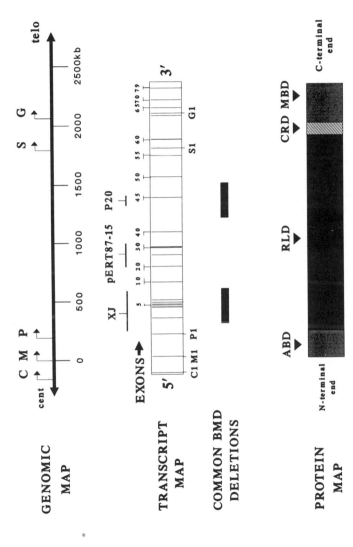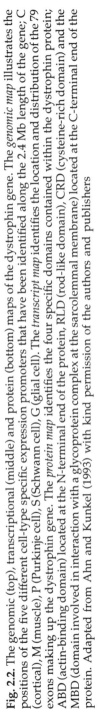

Fig. 2.2. The genomic (top), transcriptional (middle) and protein (bottom) maps of the dystrophin gene. The *genomic map* illustrates the positions of the five different cell-type specific expression promoters that have been identified along the 2.4 Mb length of the gene; C (cortical), M (muscle), P (Purkinje cell), S (Schwann cell), G (glial cell). The *transcript map* identifies the location and distribution of the 79 exons making up the dystrophin gene. The *protein map* identifies the four specific domains contained within the dystrophin protein; ABD (actin-binding domain) located at the N-terminal end of the protein, RLD (rod-like domain), CRD (cysteine-rich domain) and the MBD (domain involved in interaction with a glycoprotein complex at the sarcolemmal membrane) located at the C-terminal end of the protein. Adapted from Ahn and Kunkel (1993) with kind permission of the authors and publishers

cases; the bringing together of two such sequences from different regions of the same chromosome may permit double-strand breaks to form with illegitimate repair and consequent deletions, without an associated crossing-over event. However, Pizzuti has also found evidence that deletion break-points can lie within transposable elements that are normal constituents of intron 44 (the P20 intron). One case was found in which the distal breakpoint was located within the retrotransposon element, THE-43D (2.3 kb in size, and flanked by 350 bp long terminal repeats). One further such case, with distal breakpoint within the same transposon in intron 44, was found in 13 cases with deletions of exon 44 only. These studies conflict with the conclusions of the earlier studies, and suggest that specific sequences may indeed trigger the deletions in this intron.

The possibility that some families may have an increased risk of undergo-ing mutations at the DMD gene has been suggested as an explanation for the occurrence of several, apparently independent mutations in the same family. While some reports are likely to be coincidental, especially where the affected males have different mutations and are related through healthy males, there is a possibility that a first mutation could act as a premutation, perhaps significantly increasing the probability of a second mutation.

DUPLICATIONS

In addition to deletions, some cases of DMD and BMD have been associated with duplications within the DMD gene; like the deletions, these can be very useful in counselling if a clear junction fragment is produced. The frequency of duplication events is much less than that of deletions, estimated at 6% of all cases of DMD/BMD; however, this may be an underestimate, and the use of quantitative Southern blot analysis may detect duplications in as many as 14% of cases of DMD. Duplications can be extremely large, such as in a man with BMD in whom 400 kb of the DMD gene are duplicated, resulting in a DMD gene of over 100 exons and a protein of 690 kDa. Insertion mutations need not always be the result of tandem duplication events.

FRAMESHIFT MODEL OF DMD/BMD DIFFERENCES

The difference in severity between DMD and BMD cases has no apparent relation to the extent of DMD gene deletions. This led Monaco et al. (1988) to analyse the intron/exon border sequences in controls and in DMD/BMD cases. They suggested that DMD would arise when a mutation results in the complete absence of the gene product (dystrophin), or when an intragenic or intron-splicing mutation causes a frameshift of translation and hence prema-ture chain termination; BMD would arise when a deletion resulted in an in-frame loss of exons. They identified large BMD deletions which were compatible with the production of shortened but semi-functional protein products because the reading frame was not disrupted, whereas even a small deletion that resulted in a frameshift was associated with the DMD pheno-type.

This model was supported by the findings of other groups, but evidence did then accumulate that the model was not applicable to all deletions. Malhotra and colleagues set out to test the reading frame model by sequencing across the exon/intron boundaries of the first 10 exons; deletions of exons 3–7 'should' all result in DMD because the reading frame was shifted, but apparently identical deletions of these exons were found in six cases of BMD, five cases of intermediate severity muscular dystrophy and in three cases of DMD; some other frame-shifting deletions resulted in inter-mediate disease. It was proposed that differential splicing could restore the reading frame, or that new in-frame translational start sites could be produced downstream from the deletion and dependent upon the normal promoter, or even that a second promoter could exist downstream of the deletion in the large intron between exons 7 and 8. Exceptions to the frameshift model have also been reported in other studies; a multi-centre collaborative study found that the frameshift model accords with the clinical phenotype in 92% of cases of DMD/BMD (Koenig *et al.*, 1989). The pattern of deletions observed suggested that there may be many individuals with in-frame deletions that are essentially asymptomatic, because in-frame deletions in some regions of the gene are not observed in known cases of muscle disease. Deletions within the DMD gene that affect solely intronic sequences are already known in normal individuals.

The result of studies of dystrophin mRNA and dystrophin protein in muscle from the anomalous cases that do not match the frameshift model are interesting. In-frame deletions with a DMD phenotype have been shown to be associated with slightly reduced levels of a truncated mRNA, but effectively nil dystrophin protein, indicating impaired translation or instability of the protein; while out-of-frame deletions with a BMD phenotype are associated with the presence of alternatively spliced, in-frame transcripts as suggested by Malhotra.

POLYMERASE CHAIN REACTION DETECTION OF DELETIONS

The sequencing of exons in the DMD gene permitted the application of polymerase chain reaction (PCR) techniques to the detection of deletions in the DMD gene. Caskey and colleagues reported a multiplex PCR cocktail that amplified sequences from six different, deletion-prone exons, and hence detected 79% of deletions; this avoided the need for Southern blot analysis of DNA from 37% of males with DMD. The detection of 98% of all deletions by means of two multiplex PCR reactions is now a standard diagnostic procedure.

It is possible to amplify the muscle-specific promoter region for the DMD gene, and mutations in this region provide an explanation for some individuals with a reduced abundance of dystrophin of normal molecular weight and without detectable exonic deletions.

Quantitative PCR techniques are being developed to identify female carriers of known deletions, and to identify males with duplications within the dystrophin gene.

RNA STUDIES OF DMD GENE MUTATIONS

One problem that is not solved by Southern blot or PCR studies of genomic DNA from cases of DMD is the identification of the DMD gene mutations in the 40% of families in which deletions or duplications are not found. Reverse transcription (RT) of total mRNA from males with DMD/BMD, with subsequent PCR amplification of the dystrophin cDNA, enables deletions and duplications to be recognized and also has the potential for revealing point mutations that may account for the disease in cases with no gross gene rearrangement. It was found that the rate of 'illegitimate' transcription of the DMD gene in leucocytes is sufficient for this system to detect deletions.

Ten overlapping sets of nested primers were utilized to cover the entire dystrophin mRNA while employing favourable PCR conditions. Deletions of the gene resulted in predictably shorter transcripts, and the predictions of the frameshift model generally conformed with the results in the 26 cases studied (Roberts *et al.*, 1991). Alterations in transcript size were identified in female carriers of known deletions, and permitted unambiguous identification of the carrier state when the mutation was known to be a deletion or duplication. Interestingly, Roberts and colleagues encountered a series of alternatively spliced transcripts specific to the rearranged genotype, which accounted for most of the clinical features that were unexpected on the frameshift model.

Applying methods of chemical mismatch detection to the amplified (and grossly normal) dystrophin cDNA derived from affected males with no apparent deletion or duplication has enabled Roberts *et al.* (1992) to identify disease-causing point mutations in seven out of seven individuals studied in this way. These have been concentrated at the 3' end of the gene, indicating that the C-terminus region is of great importance for the function of dystrophin.

A further, very promising extension of this work has been the development of the protein truncation test (Roest *et al.*, 1993). In this method, RNA is isolated, reverse-transcribed and amplified in 5–10 overlapping regions; however, the nested PCR reaction then utilizes a modified primer that incorporates a T7-promoter and a (eukaryotic) translation initiation signal. The RT-PCR products can thus be transcribed and translated *in vitro*, and the size of the protein product determined by gel electrophoresis. The region of the gene in which the DMD-causing mutation has occurred will then be identified by the reaction that results in an abnormally small protein product, and the approximate location of the mutation within that region can be inferred from the size of the abnormal protein product found. The precise mutation can then be identified by the sequencing of only a short section of the DMD gene. This method can be applied to females (to identify carriers) as well as to affected males. The mutations in all of 13 non-deletion DMD cases studied have been identified precisely in this way, and all were point mutations causing premature translation termination—as would be expected in cases of DMD.

PCR amplification of genomic DNA has also been used to identify point mutations—as with the identification of a nonsense mutation in exon 8 initially detected by single-strand conformation polymorphism (SSCP) techniques—but the use of RT-PCR to study the dystrophin mRNA seems more promising because fewer reactions are required to cover the entire gene.

DYSTROPHIN

In the mouse, the X chromosome *mdx* mutation causes a myopathy much milder than the DMD disease phenotype, but the gene is located in the portion of the mouse X chromosome homologous to the DMD gene region in humans. Before the full sequence of human dystrophin cDNA was known, antibodies were generated against the murine protein. Two portions of murine heart muscle dystrophin cDNA (corresponding to 30 and 60 kDa segments of the protein) were fused to the 3′ terminus of the *Escherichia coli* *trpE* gene, in a plasmid designed for use as an expression vector. Polyclonal antisera were then raised against the proteins produced in this fashion *in vitro*, and the cytoskeletal location of dystrophin was established. With the sequencing of the human and murine dystrophin genes, the homology of more than 90% between the human and murine amino acid sequences was demonstrated. This close homology between man and mouse dystrophin genes has proved very helpful in the elucidation of dystrophin structure and function.

With the availability of the complete sequence of human dystrophin, predicted from the full cDNA sequence, it became possible to make predictions of the likely structure and function of this molecule (Koenig *et al.*, 1987, 1988). There is a 240 amino acid N-terminal that shows homology to the actin-binding domain of α-actinin. Then there is a rod domain composed of 1200 amino acids in 25 triple-helical segments that resemble the repeat domains of spectrin (a cytoskeletal component), and which also share some immunological features with actinin. This is followed by a 280 amino acid cysteine-rich domain that resembles the C-terminus of α-actinin, and then a 420 amino acid C-terminal domain with no homology to any previously described protein (Fig. 2.2). It was thought that the C-terminus was required for binding to the sarcolemma, but this is not necessarily so; a truncated dystrophin, which lacks the C-terminal 175 kDa, is nevertheless still localized to the muscle membrane in a male with a DMD phenotype.

DYSTROPHIN ANTIBODIES

Even before the full cDNA sequence was reported, antibodies were raised to dystrophin, to permit further studies of the protein. Dystrophin was shown to be of low abundance in muscle, and it was confirmed that dystrophin was not nebulin (as had been conjectured). Immunocytochemical and immunoelectron microscopic studies localized dystrophin to the submembranous cytoskeleton and confirmed its absence in cases of DMD. The presence of dystrophin in a lattice on the cytoplasmic surface of the cytoskeleton has been demonstrated; dystrophin is concentrated, along with α-actinin, at the sites of connection of the sarcomeres to the plasma membrane, in the region of the I bands.

The implications of this knowledge for our understanding of the function of dystrophin is still not clear. Dystrophin-less muscle fibres are more susceptible to osmotic lysis, and they show increased rates of protein degradation, possibly associated with elevated calcium activity in the muscle fibres. It has been suggested that the lack of dystrophin may not directly impair the tensile strength of the membrane, but may predispose muscle fibres to shed membrane in conditions of shear stress; this would result in a lower surface/volume ratio, which would then account for the susceptibility to osmotic lysis. The association of dystrophin with membrane glycoproteins will be discussed under 'Related genes and dystrophies'.

Further studies examined the dystrophin in muscle biopsy samples from cases of DMD, BMD and other muscular dystrophies by western blotting (Hoffman et al., 1988). These authors found that biopsies from cases of DMD usually contained no dystrophin, or < 3% of normal dystrophin levels, while muscle from BMD cases often contained dystrophin of apparently abnormal (usually smaller) size but of relatively normal abundance. Other types of dystrophy had apparently normal dystrophin (a normal proportion of dystrophin : myosin). Immunofluorescence studies demonstrated an absence of dystrophin in DMD and patchy, discontinuous staining of the peripheral membrane of the myofibres in BMD (Fig. 2.3a–c); other dystrophies yielded normal results on immunoblotting and immunofluorescence studies, except for a dystrophin-negative case of Fukuyama muscular dystrophy—an autosomal recessive dystrophy with severe intellectual impairment that is found most frequently in Japan. We will return to Fukuyama dystrophy under 'Related genes and dystrophies'.

Studies of patients with BMD have established that the abundance of dystrophin is a useful guide to the individual's prognosis. As well as in-frame deletions, in-frame duplications of the DMD gene can be associated with a BMD phenotype that correlates with the findings of western blotting, giving a much enlarged dystrophin molecule of reduced abundance. Some regions of the gene must remain intact for a deletion to result in BMD rather than DMD; it appears that deletions of the C-terminus, or of the N terminus and part of the proximal rod domain, will usually result in a severe, DMD phenotype. However, the finding of a patchy dystrophin appearance on immunocytochemistry may be an artefact of specific anti-dystrophin antibodies, and continuous staining of the muscle fibre cytoskeleton can be found.

These and other studies have ensured that dystrophin analysis is now a standard part of the pathological examination of a muscle biopsy from any case of suspected or possible Xp21 muscular dystrophy. To generate clinically reliable information, however, it is important for each biopsy to be examined with at least three antibodies (specific for the N-terminal, rod and C-terminal domains), and in some cases Western blotting is necessary to determine the molecular weight and abundance of the dystrophin (Fig. 2.4).

MONOCLONAL ANTIBODIES

The polyclonal dystrophin antibodies had the disadvantage of cross-reacting to other muscle proteins, such as α-actinin and spectrin. Monoclonal antibodies to dystrophin were generated, and these were of much greater (although not always absolute) specificity.

Monoclonal antibodies have also proved useful in their specificity for defined regions of dystrophin. The introduction of transposons carrying chain termination codons into a length of dystrophin cDNA in an expression plasmid permitted the production of a set of truncated dystrophin fusion proteins, and the mapping of a set of 22 monoclonal antibodies against these targets to define the region of protein required for antibody recognition. Some antibodies require a sequence as short as 7–17 amino acids for recognition. This permits a correlation to be established between the extent of an interstitial molecular deletion and the corresponding set of antibodies which fail to recognize the resulting, low molecular weight dystrophin.

FURTHER IMMUNOCYTOCHEMICAL STUDIES OF DYSTROPHIN

There are some apparently dystrophin-positive fibres in muscle biopsies from at least 50% of boys with DMD (Fig. 2.3d). Given that many such affected boys have inherited an intragenic deletion from their mother, there would seem to be little chance that a true revertant mutation could occur. One mechanism by which such 'revertant', dystrophin-positive fibres could arise would be by 'exon-skipping', in which one or more non-deleted exons is not included in the final mRNA because of modified RNA processing; this could restore the reading frame, and hence dystrophin production, from a proportion of mRNAs (Sherratt et al., 1993). Another mechanism that could account for the 'revertant' fibres would be somatic mutations within the nuclei of the

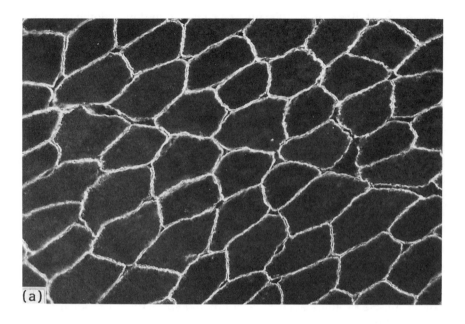

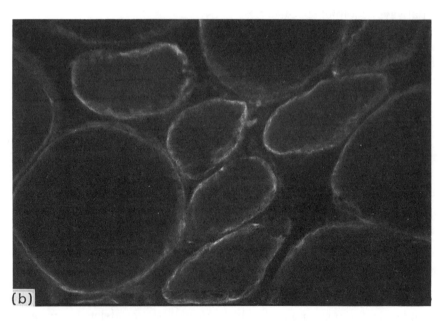

Fig. 2.3. Immunocytochemical analysis, using the anti-dystrophin monoclonal antibody Dy4/6D3, of frozen sections of skeletal muscle from a control (a), a BMD patient (b), a DMD patient (c) and a second DMD patient displaying a revertant muscle fibre (d). The fluorescent staining of the sarcolemma is complete in all muscle fibres from

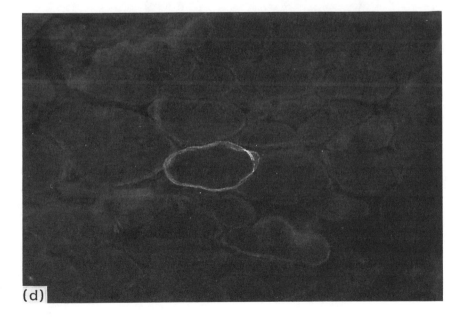

control, becomes 'patchy' in the BMD muscle and is completely absent from the DMD muscle, except for the infrequent appearance of 'revertant' muscle fibres. Kindly provided by Louise Nicholson

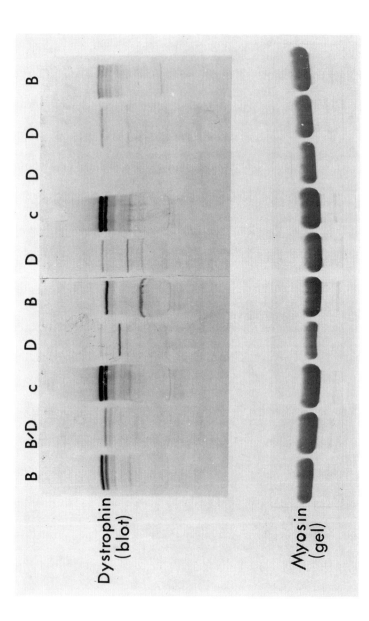

Fig. 2.4. Western blots of separated muscle proteins from two controls (c), three BMD patients (B), and four DMD patients (D) probed with the anti-dystrophin monoclonal antibody Dy4/6D3. The strongly staining dystrophin doublet present in the two control samples contrasts with the reduced staining intensity or the reduction in molecular weights of the dystrophin-staining bands in the BMD and DMD patient samples. A stain for the myosin heavy chain present in each sample controls for the quantity of muscle tissue analysed in each lane of the gel. Kindly provided by Louise Nicholson

muscle fibres (or of the myoblasts, before they become involved in regeneration), which could extend a deletion and thereby restore the reading frame. This may be more likely to account for the immunocytochemical appearances than exon-skipping, because the latter would be expected to result in a low-level but even production of dystrophin. Evidence that the 'second mutation' hypothesis accounts for 'revertant' fibres in mouse and man is accumulating; it appears that each dystrophin-positive fibre or cluster of fibres is associated with an independent second mutation, with positivity for a different set of antibodies specific to a range of epitopes. Studies on the molecular weight of the dystrophin present in biopsies from such cases are compatible with this suggestion of second mutations that restore the reading frame.

DYSTROPHIN EXPRESSION IN NON-MUSCLE TISSUES

Monaco and colleagues used Northern blots to demonstrate expression of the dystrophin gene in small intestine and lung as well as muscle, but not in brain. Nudel and colleagues employed an RNAse protection assay to detect dystrophin gene expression in mouse brain, and (in a developmentally regulated fashion) in myogenic cell cultures. When PCR techniques were used to increase the sensitivity still further, Chelly and colleagues detected low levels of transcription in a wide range of cell types, including lymphoblastoid cells.

The brain transcript and its encoded protein were compared in the rat, and found to differ in the 5' exon, suggesting the existence of tissue-specific promoters. This was confirmed in humans (Gorecki et al., 1992), and it was shown that the sequence and length of the 2 kb of transcript at the 3' end (the domain with no homology to previously recognized proteins) differs significantly in different tissues, because of differential splicing.

Much interest has focused on the brain transcripts of dystrophin. The brain promoter lies at least 90 kb 5' to the muscle promoter, and 400 kb from exon 2, to which it is spliced. Mental retardation is not a feature of two cases of DMD in which the brain promoter region has been deleted, and no evidence points to the involvement of this region specifically in cases of DMD with mental retardation (although deletions of exon 52 may be involved in this way).

It is becoming clear that the pattern of expression of dystrophin is under very complex control. In addition to the predominant 'brain' and 'muscle' transcripts, there is also a 4.8 kb apodystrophin 1 expressed in schwannoma cells; this shares exons with the 5' end of dystrophin. Furthermore, there are at least two additional transcripts found in specific cell types within the brain, and transcribed from different promoters. The first exon of the transcript found in cerebellar Purkinje cells is distinct from those of the muscle and brain dystrophins, and it is encoded within a large intron between the muscle promoter and the second exon. The relative rates of transcription from the muscle and brain promoters also differs in different cell types from within the brain (neuronal versus glial cells).

It is also clear that the expression of dystrophin is subject to developmental regulation from the finding that truncated dystrophin molecules can be found in fetal muscle studied after termination of pregnancy at 12 weeks on account of a very high (>99%) risk of DMD. Such truncated dystrophin molecules may be recognized by N-terminal antibodies and sometimes by rod domain antibodies, but not by C-terminal antibodies. At 12 weeks' gestation, dystrophin is found in the cytoplasm as well as the membrane. Unless a range of antibodies were employed, or western blotting performed as well as immunocytochemistry, there would be a risk of a 'positive' dystrophin study (i.e., the presence of dystrophin) being misinterpreted as evidence that the fetus was unaffected; these truncated dystrophin molecules would not usually be detected in post-natal life.

In summary, in addition to the three (or more) promoters for the full-length dystrophin (muscle, cortical and cerebellar Purkinje cell) with their distinct first exons, there are two C-terminal promoters, each associated with its own first exon, which encode proteins of 71 kDa (present in many tissues, including glial cells) and of 116 kDa (specific to Schwann cells) (Fig. 2.2). There is also variation between tissues in the pattern of 3' splicing, and several embryonic forms of dystrophin exist. The functional significance of this variation is not yet clear. At least 11 different isoforms have been reported in the mouse, with different patterns of expression in a variety of tissues including skeletal muscle, cardiac muscle and brain, and with distinct embryonic and adult C-termini in several tissues (Bies et al., 1992).

DYSTROPHIN IMMUNOCYTOCHEMISTRY IN FEMALES

One approach to the identification of female carriers has been to examine dystrophin production by differentiating clonal myoblasts in vitro. This typically demonstrates two populations of myoblasts—dystrophin-positive and dystrophin-negative—when the woman from whom the original muscle was obtained is a carrier of DMD. Another approach has been to examine the pattern of dystrophin distribution in muscle tissue obtained by biopsy from females who may be carriers of DMD. Immunocytochemistry performed on samples from manifesting carriers of DMD often reveals a patchy distribution of dystrophin-deficient fibres, as expected from considering the biology of X inactivation. Manifesting carriers of BMD may also show patchy abnormalities, including the patchy absence of dystrophin. However, it appears that only a minority of non-manifesting (healthy, non-myopathic) but definite carriers of DMD reveal a mosaic pattern of dystrophin deficiency.

Dystrophin studies on muscle from a t(X;12) (p21.2;q24.33) female manifesting signs of limb girdle muscular dystrophy showed that 64% of muscle fibres were dystrophin-negative, while 99.4% of leucocytes had inactivated the intact X chromosome; this girl was affected by a clinically mild myopathy. The reason why she has been spared worse effects, and why so many fibres (one-third) are dystrophin-positive, may be that the dystrophin that is produced by some nuclei is able to compensate for the lack of dystrophin mRNA from the other nuclei in the multinucleate muscle fibres. This may be the result of compensatory over-production of dystrophin, of increased stability of mRNA or protein, of the tendency for dystrophin-negative regions of fibres to undergo necrosis, or it could result if the skewing of X-inactivation apparent in leucocytes is not representative of muscle nuclei. If the compensation or the stability model is correct, then this may account for the rarity of clinically manifesting carriers of DMD among those with no cytogenetic aberration.

Finally, it has become clear that females with a limb girdle muscular dystrophy but no family history of DMD may be manifesting carriers. This can often be demonstrated by dystrophin immunocytochemistry—at least 10% of such women with a creatine kinase level of >1000 IU/l are shown to be DMD carriers in this way (Hoffman et al., 1992). This issue is discussed further under 'Related genes and dystrophies'.

CLINICAL APPLICATIONS OF THE GENETIC STUDIES

IDENTIFICATION OF FEMALE CARRIERS BY LINKAGE STUDIES

Traditional methods of calculating reproductive risks in DMD families have entailed pedigree analysis with the alteration of risks in the light of serum creatine kinase (CK) values in possible carrier females. This enzyme is released from the muscle of affected boys, so that their serum levels are greatly raised; serum levels do tend to be above average in carrier females, but there is a considerable overlap between the CK values of carrier women and the normal population, so that an odds ratio has to be employed to relate the CK value to the likelihood of a woman being a carrier of DMD. This likelihood, combined with the woman's position in the family—her relationships particularly to the affected male(s) and to other unaffected males—allows her risk of being a carrier to be calculated.

The use of early DNA markers linked to DMD allowed women to be given additional information about the risk of their carrying DMD, if their family structure permitted a linkage analysis (it is usually necessary for at least one affected male to be sampled). The ease and accuracy of these calculations have improved as more polymorphisms have become available, particularly with closer markers of greater heterozygosity and therefore of greater informativeness.

The finding that the flanking markers had recombination rates of up to 15%, whereas the pERT and XJ markers' recombination fraction was around

5% (Kunkel et al., 1986), emphasized the importance of finding further, more highly polymorphic, intragenic markers. With such high recombination rates, the combined use of intragenic and flanking markers was bound to lead to the termination of many healthy fetuses in whom at least one flanking marker had recombined with the DMD gene. The high recombination rates between the original flanking markers and DMD also meant that these markers were virtually useless in families where phase was uncertain; i.e., where it is only inferred that a woman carries the DMD mutation on the same X as one of the two marker alleles, this inference may well be incorrect if the recombination rate between marker and disease is not very small.

Experience gained over decades has shown that the reproductive risks given to families have largely been accurate, justifying the assumptions on which the risk calculations were performed, and that the more recent application of DNA methods to carrier identification has greatly clarified the situation of many of those women for whom it had not been resolved by the more traditional methods of carrier testing.

Recombination within the DMD gene is known to occur with greater frequency than in many gene loci; for example, the recombination rate within DXS164, between pERT87-1 and 87-15, is about 4%, suggesting the presence of a recombination hot spot. Estimates of recombination across the whole DMD gene became possible with the development of PCR-based polymorphism detection methods which could identify very closely flanking $(CA)_n$ micro-satellite repeat polymorphisms, so that 12% recombination across the gene now appears to be a reasonable estimate. This corresponds well to the 5% recombination between the pERT markers and the disease, but is some five times higher than would be expected across the physical distance of 2.3 Mb.

The same factor(s) that lead to the generation of deletions in particular regions within the gene may also lead to the high recombination rate, but it is not known whether the recombination occurs in just a few hot spots, or whether there is a diffuse increase in recombination across the whole gene.

IDENTIFICATION OF CARRIERS BY OTHER MOLECULAR METHODS

The direct detection of the DMD-causing mutation is the ideal form of carrier detection. In a few families (perhaps 1%) a junctional fragment is recognized on Southern blotting by one of the standard cDNA markers. Many deletion carriers may be directly identifiable using large fragment gel electrophoresis (see 'Physical mapping of the DMD gene region', above), although this has not proved popular for diagnostic work because of the need to process the blood sample rapidly after venepuncture, and because of the technically difficult PFGE that is required. Similarly, dosage-based (Southern blot or PCR) and RNA-based methods of carrier detection have been referred to in previous sections ('PCR detection of deletions; and 'RNA studies of DMD gene mutations'), although these technique have not as yet proved sufficiently robust to be widely applied in diagnostic laboratories.

More promising approaches to carrier detection for diagnostic application are now emerging. It is possible to use highly informative polymorphisms within regions that are commonly deleted to identify females who are hemizygous for the deleted region, and must therefore be carriers. It is also possible to use single-strand conformation polymorphism analysis to detect deleted regions (and potentially point mutations too) in carriers, and cytogenetic analysis with *in situ* hybridization using cosmid clones from commonly deleted introns within the DMD gene can also be used to identify carriers, who will have only a single signal per nucleus for cosmids derived from the region deleted in their family. The refinement of these promising techniques is likely to lead to further improvement in the information provided to females in families in which a deletion has been identified.

For families in which the affected male has no identifiable exonic deletion, and for families in which no affected male survives to provide a sample of DNA, the protein truncation test is likely to become established as the method of choice for identifying female carriers. This has already been discussed under 'RNA studies of DMD gene mutations'.

NEWBORN SCREENING FOR DMD

Because the mothers of boys suffering from DMD are frequently carriers of the disease, and because the condition is not usually diagnosed until the boy is several years old, it is possible for the mother of an isolated case (the first affected boy in a family) to conceive a second affected son before the first boy is diagnosed. In such a family, two—or even more—boys may be diagnosed simultaneously. This causes great distress, and may also result in anger and bitterness in the family, especially if they have suspected some medical problem in the first affected boy, but the various health professionals have failed to identify him as having DMD. The average diagnostic delay between initial concern and formal diagnosis can be more than 2 years, and the average age at diagnosis in some series has been as late as 5–6 years.

Because of such experiences, a system of CK screening for boys who were not walking independently at 18 months of age was proposed. This should identify about 50% of boys with DMD, who might not otherwise be diagnosed until 5+ years. Such a screening programme was established and evaluated in Wales; while some affected boys were identified, too many logistic and administrative difficulties arose to justify the extension of the scheme.

The failure of the selective 18-month screening programme, coupled with the advent of molecular genetic methods which have greatly improved the accuracy of genetic counselling for DMD, led our group in Wales to establish and to evaluate a newborn screening programme for DMD. Newborn screening had previously been avoided because of the possibility of causing serious emotional trauma; however, the families of affected boys were enthusiastic about newborn screening, and 94% of mothers of newborn infants in a maternity hospital declared (hypothetically) that they would accept newborn screening for DMD if it were offered. It was already known that such screening is technically feasible, and that other causes of elevated serum CK in boys, beyond the first few days of life, are few. The question of whether the families are too greatly distressed or disrupted by the experience of an early diagnosis, or whether they find this preferable to a delayed diagnosis, is being addressed by a careful social evaluation of our pro-gramme; preliminary experience has been favourable (Bradley *et al.*, 1993).

TOWARDS GENE THERAPY?

There are several major obstacles to successful gene therapy for DMD. The dystrophin gene is the largest known (2.3 Mb), and the complete transcript (at 14 kb) is too large for incorporation into retroviral vectors. The gene is expressed not only in striated and cardiac muscle, but also in many other tissues, including the brain. The relevant tissues cannot be removed from the body for treatment (as for bone marrow) or simply injected; it is difficult to think of a viral vector that would readily target all the correct tissues, or that could be constructed to do so.

One therapeutic avenue that has been explored is the transfer of donor myoblasts into dystrophic muscle. The injection of 8 million myoblasts from his father into a small muscle of the foot of a boy with DMD resulted in the appearance of dystrophin-positive fibres in the injected muscle. However, the results of subsequent, more ambitious myoblast transfer experiments by the same group of workers and by others have been very disappointing; the massive research effort involved has yielded little of clinical value to affected boys. DNA can also be injected directly into muscle cells and be capable of functioning, but this may have many of the practical problems of myoblast therapy.

Despite these difficulties, there has been considerable progress towards some form of rational gene therapy for DMD. The realization that the truncated dystrophin molecules found in certain men with mild forms of BMD can still function very satisfactorily has prompted the suggestion that such a dystrophin 'minigene' could be packaged into a viral vector much more readily than the full-length cDNA. Thus, a deletion of 46% of the transcript results in the production of a dystrophin of 220 kDa, and is compatible with only a very mild BMD phenotype—with ambulation into the seventh decade. Such dystrophin 'minigenes' have been transferred into *mdx* (the murine DMD homologue) mice *in vivo* using retrovirus mediated and adenovirus-mediated techniques, and the latter has achieved long-term correction of the mouse dystrophy (Vincent *et al.*, 1993).

Another advance has been the synthesis and expression of a full-length (14 kb) cDNA for (mouse) dystrophin, using lambda phage inserts from cDNA libraries constructed with dystrophin-specific primers. The complete cDNA was cloned in a Bluescript vector for sequencing, and then into a mammalian SV40 expression system which was transfected into COS cells.

On a different tack, an advance that has been achieved independently by at least three different groups is the construction of a YAC contig spanning the entire human dystrophin gene. By mating YAC-containing haploid yeasts of different mating types and then selecting appropriate products of homologous meiotic recombination, it should be possible to construct a single YAC containing the entire DMD gene. This will greatly facilitate the improved understanding of DMD gene structure and expression, and may facilitate the development of useful therapies.

RATES OF MUTATION WITHIN THE DUCHENNE GENE

MALE AND FEMALE MUTATION RATES

Risk calculations for carrier status for mothers and other female relatives of boys with DMD have been carried out on the assumption that the mutation rate at the DMD locus is equal in males and females. If this is true, then the mother of a boy with DMD has a two-thirds chance of being a carrier; conversely, the proportion of affected boys whose disease has arisen by new mutation is one-third. These conclusions were established by Haldane, and are of more than theoretical importance, because the accuracy of genetic counselling depends upon them. If the mutation rate is higher in males, then more mothers of isolated affected boys will be carriers, but fewer females in the extended family. If the mutation rate is higher in females, then fewer mothers will be carriers.

The mutation rate for DMD has been measured in a number of populations, with results of the order of 10^{-4} per generation. While there is no firm evidence of differences in the rate of mutation between different families or different ethnic groups, there have been suggestions that the mutation rate may differ in these ways. The number of boys with DMD in one region of England includes an excess of those of Indian origin, and a rather low number of those of Pakistani origin. It has been suggested that the DMD genes of the Indian group may contain repetitive elements that predispose to unequal recombination, or some transposable element that can induce mutations. In this context, it is interesting that parents of mixed racial origin may have an excess of DMD boys, and that the Pakistani group with an apparently low incidence of DMD tends to find marriage partners within the extended family group. In *Drosophila*, transposable element-induced mutations are much more frequent in crosses between different stocks of flies; inbreeding may provide some protection against transposable element-induced disease. The finding in Brazil of four DMD families with affected individuals related through paternal lines could also be seen as indicating a transmissible tendency to undergo DMD gene mutation.

Putting to one side these suggestions of differences in mutation rate between groups and families, we should return to consider the possibility of differences in mutation rate between the sexes, because this is of more immediate applicability to counselling and to the interpretation of molecular studies. Some deviations from the equality of mutation rates in the two sexes has been reported, with an excess of new mutation cases (higher female mutation rate) reported from one centre and the opposite from others. The use of molecular genetic analysis to measure the sex ratio of mutation rates was proposed by Muller and Grimm and by Karel et al. (1986); Karel et al. point out that bias in the ascertainment of families may be an important confounding factor, but that it can be avoided through selecting families with at least one healthy boy at the time of ascertainment of the single affected boy, who must have inherited the DMD gene haplotype from his maternal grandfather; determining the proportion of healthy brothers with the same origin of their DMD gene as in the affected boys allows the sex ratio of mutation rates to be calculated without bias.

Large numbers of families must be analysed to achieve an adequate power to detect a deviation from a sex ratio of unity. One study found no evidence of a deviation from equality, while another study found a ratio of grandpaternal to grandmaternal origin of deletion mutations was 32:49 (van Essen et al., 1992). This study included deletion cases only, in which the mother had been shown by dosage to be the first carrier in the family.

MOSAICISM

Anomalous results generated in some family studies led to the recognition of mosaicism for the DMD mutation: some mutations must have occurred at a mitotic cell division. For example, a female heterozygous for an intragenic polymorphism may be shown to have transmitted a deletion of the relevant marker to more than one child (of either sex). Cases of male mosaicism have also been reported, as when two of five daughters in one family carried a DMD gene deletion derived from their unaffected father's X chromosome—the result of grandpaternal germinal mosaicism. Such phenomena are important, and (if unrecognized) can lead to errors in prenatal diagnostic testing.

Several studies have attempted to measure the extent of the reproductive risk resulting from mosaicism in the mothers of sporadic cases of DMD. Passos-Bueno compared the CK levels in true obligate carriers of DMD (women with an affected son and brother) with the CK levels in probable carriers (women with at least two sons, or one son and one carrier daughter). Higher CK levels were found in the truly obligate carriers, and it was estimated that 12% of the probable carriers were in fact mosaic for the mutation.

Haplotype analysis has been employed to determine the parental origin of new deletion or duplication mutations at the DMD locus (where the mutation was absent from the lymphocytes of the mother or of the grandparents). The recurrence risk for the high-risk haplotype among the future children of apparently non-carrier mothers of an affected boy was 14% (7% if haplotype information is not available). The multicentre study of van Essen *et al.* (1992) estimated the risk of recurrence in children with the same haplotype as that of the affected boy in families where the mother is not a carrier of the deletion or duplication found in the boy. The risk is given as 20% (95% confidence interval 10–31%), or 10% risk to future males if the haplotype information is not available. Preliminary data also suggest that the risk of mosaicism may be greater for proximal than for distal deletions. Further clarification of the rates of mosaicism in families with deletion, duplication and point mutations will be important for accurate counselling.

The possibility that the origin of DMD gene deletions or duplications might be related to the occurrence of recombination events at this locus was proposed by Winter and Pembrey in 1982. Some deletion events have indeed been found to be associated with recombination events between flanking markers, but more work will be needed with the new, close flanking and intragenic markers to examine this relationship in more detail. The group of Worton in Toronto (Hu *et al.*, 1990) have established that unequal sister chromatid exchange (SCE) is the likely cause of certain DMD-causing gene duplication events, found in about 6% of cases of DMD. Unequal SCE is a mechanism that can occur in either sex, whereas unequal non-sister chromatid exchange, like unequal recombination events, can only occur in females. The fact that a grandpaternal origin of unequal SCE-derived mutations has been found more frequently than a grandmaternal origin, and predominantly for duplication events, suggests that other mechanisms may well apply in females and possibly with deletion events in either sex.

THE DUCHENNE GENE IN FEMALES

The identification of female carriers of DMD by molecular genetic means, and dystrophin studies in female carriers, have been discussed in earlier sections. This section addresses the question of how a mutation in the DMD gene may affect a female carrier.

CONVENTIONAL MEANS OF IDENTIFYING FEMALE CARRIERS OF DMD

The experience of carrier detection at a number of centres dates back at least 30 years. The investigations performed on possible carrier females in an attempt to clarify their carrier status have included serum biochemistry, electromyography (EMG), muscle histology and various forms of muscle imaging. By far the most useful single test that has emerged is the assay in serum of muscle enzymes, particularly CK. Combining the CK assay with pyruvate kinase assay may add modestly to the accuracy of risk determinations, but the essential step in such calculations is the incorporation of Bayesian probability to combine prior (pedigree) risks with conditional information such as CK levels. The CK result is used to modify risks on a continuous scale of the likelihood of being a carrier given the average value over three occasions; a simple cut-off value is not employed, because that ignores the overlap between normal and carrier values and prevents the essential step of incorporating the pedigree information. Even with current molecular genetic methods, CK assay and risk calculation are required when the mutation cannot be detected directly, and particularly for assessment of the status of the mothers and sisters of isolated cases of DMD. For BMD there is less experience, but likelihood tables relating risk to serum CK are available from some laboratories.

Muscle histology and EMG studies may reveal subtle abnormalities in a proportion of carriers, but are not sufficiently discriminatory for use in genetic counselling. Ultrasound studies, X-ray computed tomographic (CT) scanning and red blood cell morphology (in particular the susceptibility of erythrocytes to deformation in the presence of 1-α-lysophosphatidylcholine) may also show subtle abnormalities in female carriers, but these techniques also have not proved generally applicable in clinical practice.

MANIFESTING FEMALE CARRIERS OF DMD

Early reports of clinical abnormalities in female carriers of DMD with normal chromosomes focused especially on the heart. Emery reported that female carriers, like affected boys, have a feature in common on the ECG: the algebraic sum of the R and S waves in pre-cordial lead V1 is increased. Emery suggested that such carriers may have a latent cardiomyopathy and a predisposition to cardiac failure. These ECG findings in carriers have been confirmed, and it has become clear that a small proportion of otherwise unaffected carriers (as well as of manifesting carriers) do develop a dilated cardiomyopathy as a complication of the DMD disease process. There have also been reports of an increased incidence of mitral valve prolapse in affected males and carrier females.

Manifesting female carriers of DMD can develop symptoms at any stage from early childhood to middle adult life, frequently presenting with waddling gait, muscle cramps, calf hyptertrophy or weakness in the legs. There is frequently asymmetry of muscle bulk or strength, and the weakness is often progressive; this condition appears to have a similar incidence to autosomal recessive limb girdle muscular dystropy in females (about

1–2: 100 000). One estimate of the proportion of female carriers who manifest signs of the dystrophy is around 2.5%, although some estimates of the proportion who have features of the cardiomyopathy on investigation exceed 50% (Comi *et al.*, 1992). Dystrophin immunocytochemistry performed on biopsies from such women are usually abnormal. Female carriers of BMD may also occasionally manifest clinical symptoms of muscle weakness, usually later and more mildly than in affected males.

MONOZYGOTIC (MZ) TWINS AND DMD

Several reports have appeared of monozygotic female twins with one member of each pair suffering from a DMD-like muscular dystrophy. The most likely explanation of such situations would be that the twins are both carriers of DMD, and that they differ in the pattern of X inactivation. An autosomal recessive disorder would be expected to affect both girls equally. Burn *et al.* (1986) have suggested that grouping of cells with the same pattern of X inactivation in the early embryo may actually induce twinning, and the pattern of lymphocyte X inactivation in the two girls in their study is compatible with this. However, apparent skewing of X inactivation is not infrequent and can vary between tissues, and it would be very difficult to prove that simple chance could not account for these observations. Certainly, in other similar MZ twin pairs, the severity of disease in the affected twin has been unusually mild, and there has been some evidence of the carrier state in the clinically healthy twin.

Another possible explanation for differences between members of an MZ twin pair with respect to X chromosome gene defects has been proposed by Nance. Possibly because of breaks in the zona pellucida, the inner cell mass may split into two; such splitting may well be asymmetrical, and the smaller clump of cells may be liable to skewed X inactivation because it consists of so few cells (perhaps just a single cell). Nance predicted that some disease-manifesting twins would have unaffected twins with an opposite pattern of X inactivation from the affected twin, while others of the unaffected twins would have random X inactivation, when the splitting had been asymmetrical. Where the twinning had been asymmetrical, one would expect the size of each patch of cells showing the same X inactivation pattern in a given tissue (such as muscle) to be larger in the twin resulting from the smaller number of cells, because of the greater catch-up growth required by that twin—an intrapair difference in the size of 'spots' (as observed in the report of Burn *et al.*, 1986). This is a more satisfying explanation than the suggestion that the preferential aggregation of cells bearing the same active X can lead to twinning, and has received some experimental support (Lupski *et al.*, 1991).

RELATED GENES AND DYSTROPHIES

CLARIFICATION OF DIFFERENTIAL DIAGNOSIS

Modern molecular genetic techniques can be applied both to clarify the differential diagnosis of neuromuscular disease in the individual case, and to rationalize the classification of this important group of disorders. We have experience of males diagnosed clinically as affected by spinal muscular atrophy, who were later shown to have BMD when a dystrophin gene deletion was found in their DNA. Similarly, the diagnosis of Kugelberg–Welander spinal muscular atrophy in two sisters has been overturned with the finding of a deletion of DXS206 (XJ1-1) in both of them (on the grandpaternally derived X chromosome). That these females are manifesting heterozygotes of DMD only came to light because their half-brother was diagnosed as suffering from DMD, and he was found to have the same molecular deletion.

Molecular genetic studies of the DMD gene are also helpful in distinguishing BMD from other causes of limb girdle muscular dystrophy (LGMD) in adult males, so that cDNA and dystrophin immunocytochemistry studies on muscle are now essential in the investigation of any such patient. The high proportion of males (10/11) amongst adults with undiagnosed (non-specific) LGMD in at least one population suggests that Xp21 dystrophy may still be an important and under-recognized cause.

Manifesting female carriers of DMD or BMD presenting with features of a LGMD may not be recognized as suffering from Xp21 dystrophy unless or until they have an affected male relative, or unless they are found to have a cytogenetic abnormality. The report of Hoffman *et al.* (1992) indicates that perhaps 10% of females with myopathy and elevated serum CK have abnormalities of the dystrophin in their muscle biopsies. Earlier reports of females with muscular dystrophy will all need to be re-evaluated in the light of molecular genetic and dystrophin immunocytochemical studies.

AUTOSOMAL RECESSIVE LIMB GIRDLE MUSCULAR DYSTROPHY

While there are undoubtedly autosomal recessive forms of muscular dystrophy in childhood, found as homogeneous entities in communities in Sudan, Tunisia, Reunion Island and elsewhere, it is uncertain how common or how homogeneous this is in Europe. With the availability of molecular genetic techniques, unusual families with girls, or with girls and boys, affected by muscular dystrophy have been studied carefully. Some cases of DMD have been identified in this way, but other families have been found in which affected sibs of both sexes have opposite DMD gene haplotypes from their mother (have inherited the opposite maternal alleles), with one affected sib sharing their allele with an unaffected brother—the disorder in these families cannot be an Xp21 dystrophy.

Some families in which affected males have a DMD-like disorder, but in which X-linkage is not evident from the pedigree, may have an autosomal recessive (AR) dystrophy; the proportion of such families in which this applies may be assessed by considering the structure of families in which there is an affected girl, using classical segregation analysis. Zatz and colleagues considered 20 such families, and estimated that the 'DMD' was AR in 6.8% of cases; 2.5–4% of isolated male patients with 'DMD' would then have an AR dystrophy. Fifty isolated males from this population were tested by molecular methods (screening for cDNA deletions) and by CK testing of female relatives, and the frequency of AR DMD-like dystrophy was cautiously estimated as being 8–12%. This estimate applies specifically to this Brazilian population, and cannot be generalized. Dystrophin studies demonstrated normal appearances in one boy with gross histopathological features of dystrophy, confirming the AR inheritance in his (consanguineous) family.

The severe, Tunisian form of AR dystrophy is known to map to the pericentromeric region of chromosome 13q, while the less severe forms of AR dystrophy in Reunion Island and among the Amish in Indiana, USA, maps to chromosome 15q. Families of non-Xp21 dystrophy in Brazil exhibit heterogeneity, with a minority of the less severe families exhibiting linkage to the 15q locus, and the rest (including the families with the more severe dystrophy) not showing linkage to either 15q or to the region of the dominant LGMD locus on chromosome 5q.

Where the differential diagnosis lies between BMD and AR limb girdle muscular dystrophy (LGMD) in a single adult male, it is possible to calculate the relative probabilities of his having BMD or LGMD given assumptions about the mutation rate in males and females for BMD, the fertility of males with BMD and the incidence of AR LGMD, and given the CK levels of related females and DNA haplotype information of the affected male and other family members. Now, however, molecular genetic studies and/or dystrophin immunocytochemistry on muscle biopsy material can clarify the diagnosis much more directly.

GENES, PROTEINS AND DYSTROPHIES RELATED TO DYSTROPHIN/DMD

While isolating cDNA clones from a fetal muscle cDNA library with the intention of finding the full coding sequence of the DMD gene, Love et al. (1989) identified a cDNA clone that showed only limited homology to dystrophin cDNA sequences. Taking the non-homologous fragment of this clone and hybridizing it to a northern blot of human fetal muscle cells detected a 13 kb transcript, distinct from the dystrophin transcript. This derived from chromosome 6q, and a portion of open reading frame corresponding to 490 amino acids showed 73% (nucleotide) homology to the

C-terminal domain of dystrophin; there is conservation of the translated and untranslated portions of the transcript between a number of species.

The chromosome 6 homologue of dystrophin is localized particularly to the neuromuscular and myotendinous junctions, and to peripheral nerves and vasculature of skeletal muscle; it is expressed at increased levels in embryonic tissue and in regenerating fibres. Dystrophin antibodies that cross-react with the homologue may give misleading results because of this. Furthermore, in DMD muscle biopsies (but not in normal muscle) continuous sub-sarcolemmal staining is produced by antibodies specific to the homologue as well as by antibodies to the regions of dystrophin conserved between man and chicken. The wide tissue distribution of the chromosome 6 homologue may arise because of its presence in vascular smooth muscle; it can also be produced in transformed cells of glial and Schwann cell origin.

Because there is potential for terminological confusion in this area, we recommend adherence to the convention of the Oxford group (as presented by Suthers, reported in Clarke, 1992). The term dystrophin-related protein (or polypeptide) is confined to Xp21 variants, such as the alternate transcripts and splicing isoforms described above under 'Dystrophin'. Any non-Xp21 genes with homology to dystrophin will be termed DMD-like or dystrophin-like, and the specific chromosome 6 homologue is termed 'utrophin' because of its ubiquitous tissue distribution. No disease state has yet been identified as resulting from an abnormality of the utrophin gene.

By careful solubilization of skeletal muscle membranes, the group of Campbell have isolated dystrophin molecules in association with four glycoprotein molecules (at 156kDa, 50kDa, 43kDa and 35kDa)—the dystrophin complex (Fig. 2.5). This complex is partly embedded in the membrane. The largest of the glycoproteins (156kDa) is greatly reduced in abundance in DMD muscle. Further work has shown that the 156kDa glycoprotein and the 43kDa glycoprotein are encoded on a single mRNA, and that the 156kDa dystrophin-associated glycoprotein (DAG) is a laminin-binding glycoprotein that provides linkage between the sarcolemma and the extracellular matrix (Matsumara and Campbell, 1994). The DAGs have been examined in muscle from patients with a number of different neuromuscular disorders, and a deficiency of the 50kDa DAG has been detected in several cases of the severe, DMD-severity AR LGMD of childhood prevalent in North Africa (Matsumura et al., 1992).

It is also possible that the product of the gene responsible for the (AR) Fukuyama congenital muscular dystrophy (FCMD) and dystrophin may interact; three of 23 males with apparent FCMD were found to have deletions of the DMD gene, and Beggs et al. (1992) suggest that males with dystrophin deficiency will present with the more severe clinical features of FCMD if they are heterozygous for that condition. The relation between the DAGs and FCMD remains to be elucidated (Arahata et al., 1993).

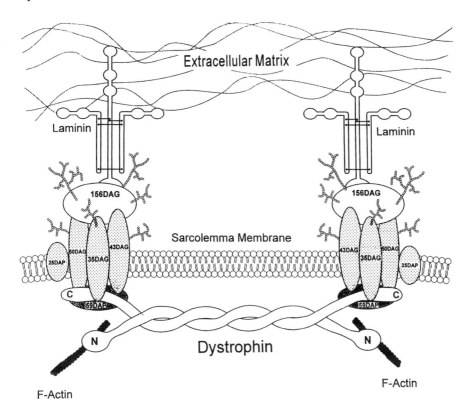

Fig. 2.5. The proposed interaction of a dystrophin homopolymer (containing either two or four dystrophin molecules) with the large glycoprotein complex functioning as a transsarcolemmal linker between actin elements, in the subsarcolemmal cytoskeleton, and laminin proteins, in the extracellular matrix. From Matsumura and Campbell (1994), reproduced by kind permission of the authors and publishers

AUTOSOMAL DOMINANT LGMD

We have already referred to the importance of dystrophin and molecular genetic studies in the clarification of the differential diagnosis of LGMD; one recent report details a father and son with clinically typical 'BMD' who in fact have an autosomal dominant (AD) LGMD. One large North American family with AD LGMD has been described with onset of symptoms at a mean age of 27 years, and a slowly progressive course; genetic linkage has been established to a region of chromosome 5q that contains the Marfan syndrome fibrillin gene and which may show homology to the region of 15q that includes the AR LGMD mentioned in the previous subsection and another fibrillin locus. Other less common forms of AD LGMD have also been described clinically (see Clarke, 1992).

SEX-LINKED MUSCLE DISEASES

In addition to DMD/BMD, the X chromosome carries loci for one other muscular dystrophy and several other neuromuscular disorders.

(1) Confusion between the phenotypes of BMD and Emery–Dreifuss muscular dystrophy (EDMD) originally led to the 'discovery' of linkage between BMD and colour-blindness. The finding that BMD is allelic to DMD has allowed work on the much less prevalent EDMD to proceed. Clinically, the condition is characterized by the early onset of contractures at large joints and of the vertebral column (often before the onset of frank muscle weakness), a slowly progressive weakness and wasting of initially humeroperoneal distribution, and a cardiomyopathy with disturbance of cardiac conduction. The disorder has been localized by linkage studies and physical mapping to a small region within Xq28, and the gene has very recently been isolated.

(2) Bulbospinal muscular atrophy (Kennedy's disease), which is caused by an expansion in a nucleotide triplet repeat within exon 1 of the androgen receptor gene on proximal Xq.

(3) X-linked Charcot–Marie–Tooth disease (an uncommon form of hereditary motor-sensory peripheral neuropathy) is associated with mutations in the connexin gene, also on proximal Xq.

(4) There is a myopathy associated with the McLeod red cell phenotype. The McLeod syndrome warrants some attention, because it maps to Xp21, as does DMD. The McLeod red cell phenotype is caused by the failure of presentation at the cell surface of the Kell blood group antigen, resulting in an abnormal erythrocyte morphology (acanthocytosis); it can be associated, as a contiguous gene deletion syndrome, with chronic granulomatous disease, DMD and retinitis pigmentosa, as in 'BB'. Without evidence of DMD, however, the McLeod syndrome includes the elevation of the serum CK and a mild, non-specific myopathy. One male has been reported with an elevated CK and a myopathic muscle biopsy but no muscle symptoms at 40 years. Dystrophin was normal in his biopsy, and the 5' end of the DMD gene was intact (excluding a major deletion of the McLeod locus extending into the DMD gene). Thus, there are probably at least two distinct forms of Xp21 myopathy.

(5) Finally, there is a rare X-linked cardioskeletal myopathy associated with neutropenia and abnormal mitochondria (Barth syndrome), the gene for which is located at Xq28.

REFERENCES

Ahn AH and Kunkel LM (1993) The structural and functional diversity of dystrophin. *Nature Genet.*, **3**, 283–291.

Arahata K, Hayashi YK, Mizuno Y, Yoshida M and Ozawa E (1993) Dystrophin-associated glycoprotein and dystrophin co-localisation at sarcolemma in Fukuyama congenital muscular dystrophy. *Lancet*, **342**, 623.

Beggs AH, Neumann PE, Arahata K *et al.* (1992) Possible influences on the expression of X chromosome-linked dystrophin abnormalities by heterozygosity for autosomal recessive Fukuyama congenital muscular dystrophy. *Proc. Natl Acad. Sci. USA*, **89**, 623–627.

Bies RD, Phelps SF, Cortez MD *et al.* (1992) Human and murine dystrophin mRNA transcripts are differentially expressed during skeletal muscle, heart and brain development. *Nucleic Acids Res.*, **20**, 1725–1731.

Boyd Y, Buckle VJ, Holt S *et al.* (1986) Muscular dystrophy in girls with X;autosome translocations. *J. Med. Genet.*, **23** 484–490.

Bradley DM, Parsons EP and Clarke A (1993) Experience with screening newborns for Duchenne muscular dystrophy in Wales. *Br. Med. J.*, **306**, 357–360.

Burghes AHM, Logan C, Hu X *et al.* (1987) A cDNA clone from the Duchenne/Becker muscular dystrophy gene. *Nature*, **328**, 434–437.

Burn J, Povey S, Boyd Y *et al.* (1986) Duchenne muscular dystrophy in one of 2 monozygotic twin girls. *J. Med. Genet.*, **23**, 494–500.

Clarke A (1992) Conference Report: ENMC Workshop on the limb-girdle muscular dystrophies. *J. Med. Genet.*, **29**, 753–755.

Comi LI, Nigro G, Politano L, Petretta VR (1992) The cardiomyopathy of Duchenne/Becker consultands. *Int. J. Cardiol.*, **34**, 297–305.

Gorecki DC, Monaco AP, Derry JMJ *et al.* (1992) Expression of four alternative dystrophin transcripts in brain regions regulated by different promoters. *Hum. Mol. Genet.*, **1**, 505–510.

Hoffman EP, Fischbeck K, Brown RH *et al.* (1988) Characterization of dystrophin in muscle-biopsy specimens from patients with Duchenne's or Becker's muscular dystrophy. *N. Engl. J. Med.*, **318**, 1363–1368.

Hoffman EP, Arahata K, Minetti C (1992) Dystrinopathy in isolated cases of myopathy in females. *Neurology*, **42**, 967–975.

Hu X, Ray PN, Murphy EG, Thompson MW, Worton RG (1990) Duplicational mutation at the Duchenne muscular dystrophy locus: its frequency, distribution, origin and phenotype-genotype correlation. *Am. J. Hum. Genet.*, **46**, 682–693.

Karel ER, te Meerman GJ and ten Kate LP (1986) On the power to detect differences between male and female mutation rates for Duchenne muscular dystrophy, using classical segregation analysis and restriction fragment length polymorphisms. *Am. J. Hum. Genet.* **38**, 827–840.

Koenig M, Hofman EP, Bertelson CJ *et al.* (1987). Complete cloning of the Duchenne muscular dystrophy (DMD) cDNA and preliminary genomic organisation of the DMD gene in normal and affected individuals. *Cell*, **50**, 509–517.

Koenig M, Monaco AP and Kunkel LM (1988) The complete sequence of dystrophin predicts a rod-shaped cytoskeletal protein. *Cell*, **53**, 219–228.

Koenig M, Beggs AH, Moyer M *et al.* (1989) The molecular basis for Duchenne versus Becker muscular dystrophy: correlation of severity with type of deletion. *Am. J. Hum. Genet.*, **45**, 498–506.

Kunkel LM and 75 co-authors (1986) Analysis of deletions in DNA from patients with Duchenne and Becker muscular dystrophy. *Nature*, **322**, 73–77.

Love DR, Hill DF, Dickson G *et al.* (1989) An autosomal transcript in skeletal muscle with homology to dystrophin. *Nature*, **339**, 55–58.

Lupski JR, Garcia CA, Zoghbi HY, Hoffman EP and Fenwick RG (1991) Discordance of muscular dystrophy in monozygotic twins: evidence supporting asymmetric splitting of the inner cell mass in manifesting carrier of Duchenne dystrophy. *Am. J. Med. Genet.*, **40**, 354–364.

Matsumura K and Campbell KP (1994) Dystrophin–glycoprotein complex: its role in the molecular pathogenesis of muscular dystrophies. *Muscle Nerve*, **17**, 2–15.

Matsumura K, Tome FMS, Collin H *et al.* (1992) Deficiency of the 50K dystrophin-associated glycoprotein in severe childhood autosomal recessive muscular dystrophy. *Nature*, **359**, 320–322.

Monaco AP, Neve RL, Colletti-Feener C *et al.* (1986) Isolation of candidate cDNAs for portions of the Duchenne muscular dystrophy gene. *Nature*, **323**, 646–650.

Monaco AP, Bertelson CJ, Liecht-Gallati S, Moser H and Kunkel LM (1988) An explanation for the phenotypic differences between patients bearing partial deletions of the DMD locus. *Genomics*, **2**, 90–95.

Ray PN, Belfall B, Duff C *et al.* (1985) Cloning of the breakpoint of an X;21 translocation associated with Duchenne muscular dystrophy. *Nature*, **318**, 672–675.

Roberts RG, Barby TFM, Manners E, Bobrow M, Bentley DR (1991) Direct detection of dystrophin gene rearrangements by analysis of dystrophin mRNA in peripheral blood lymphocytes. *Am. J. Hum. Genet.*, **49**, 298–310.

Roberts RG, Coffey AJ, Bobrow M and Bentley DR (1992) Determination of the exon structure of the distal portion of the dystrophin gene by vectorette PCR. *Genomics*, **13**, 942–950.

Roest PAM, Roberts RG, Sugino S, van Ommen G-JB, den Dunnen JT (1993) Protein truncation test (PTT) for rapid detection of translation-terminating mutations. *Hum. Mol. Genet.*, **2**, 1719–1721.

Sherratt TG, Vulliamy T, Dubowitz V, Sewry C and Strong N (1993) Exon skipping and translation in patients with frameshift deletions in the dystrophin gene. *Am. J. Hum. Genet.*, **53**, 1007–1015.

van Essen AJ, Abbs S, Baiget M *et al.* (1992) Parental origin and germline mosaicism of deletions and duplications of the dystrophin gene: a European study. *Hum. Genet.*, **88**, 249–257.

Vincent N, Ragot T, Gilgenkrantz H *et al.* (1993) Long-term correction of mouse dystrophic degeneration by adenovirus-mediated transfer of a minidystrophin gene. *Nature Genet.*, **5**, 130–134.

FURTHER READING

Emery AEH (1993) *Duchenne Muscular Dystrophy*, 2nd edn. Oxford: Oxford Medical Publications.

Nicholson LVB, Johnson MA, Bushby KMD *et al.* (1993) Integrated study of 100 patients with Xp21 linked muscular dystrophy using clinical, genetic, immunochemical, and histopathological data. Part 1. Trends across the clinical groups. Part 2. Correlations within individual patients. Part 3. Differential diagnosis and prognosis. *J. Med. Genet.*, **30**, 728–751.

3 The Cystic Fibrosis Gene: Cloning and Characterization

JEREMY P. CHEADLE and DUNCAN J. SHAW

CYSTIC FIBROSIS: THE DISEASE

Cystic fibrosis (CF) is the most common severe autosomal recessive disorder of the Caucasian population, with an estimated incidence of $1:2500$ and a heterozygote (carrier) frequency of $1:25$. Its frequency in oriental and black populations is considerably lower. Heterozygote carriers cannot be recognized clinically, but the clinical features of homozygotes for the disease gene may include meconium ileus, pancreatic dysfunction, recurrent pulmonary infections, obstructive pulmonary disease, sinus infection, nasal polyps, and sterility in male patients. An additional characteristic feature of these patients is the presence of increased sweat electrolytes, which is often used as a convenient diagnostic test.

CF is characterized by abnormal epithelial ion transport, particularly of chloride ions, in various cells including lung, sweat gland, gut and pancreas. Normally the Cl⁻ channel is activated by cAMP and associated protein kinase, but not in CF. This causes the fluid balance to be upset, resulting in dehydrated mucus which is poorly cleared, and subsequent lung infection that is the main cause of death. Affected children are treated by extensive physiotherapy to help the clearing of mucus from the lungs, and need aggressive antibiotic therapy against the potential for infection. As a result of progressive improvements in patient management, life can now be prolonged at least into the third decade for most patients.

LINKAGE STUDIES

In common with most human genetic diseases, the protein defect underlying CF was not known, and the chromosomal location of the gene had not been determined. CF was a prime candidate for the application of recombinant DNA-based linkage analysis, when these methods became available in the early 1980s.

Molecular Genetics of Human Inherited Disease. Edited by D.J. Shaw
Published 1995 by John Wiley & Sons Ltd

It was somewhat surprising, therefore, that the first linkage to CF was detected not by the new 'hi-tech' approaches, but by the use of a traditional protein polymorphism. Paraoxonase (PON) is a serum esterase enzyme, whose level of activity is polymorphic in many populations. In 1985 Eiberg *et al.* in Denmark reported linkage of PON to CF. Their studies also excluded CF and PON from large areas of the genome, but since neither marker was linked to any other locus of known chromosomal position, it was not possible to say on which chromosome the CF gene was.

The attention of those using restriction fragment length polymorphisms (RFLPs) was quickly turned to the parts of the genome that had not been excluded by Eiberg's study, and almost simultaneously three groups discovered linkage to DNA markers on the long arm of chromosome 7 (Knowlton *et al.*, 1985; White *et al.*, 1985; Wainwright *et al.*, 1985; Tsui *et al.*, 1985).

At this point it became possible for many more groups to become involved in the study, by testing the linked RFLPs on their own family material. By pooling and jointly analysing large amounts of such data, it was possible to confirm the linkage beyond any doubt and to show that a single gene appeared to be responsible for all CF. An important consequence of this work was the possibility of using the DNA probes for pre-natal diagnosis in families where a child affected with CF had already been born. At this stage, however, it was not possible to detect carriers of the CF gene since the polymorphic variants associated with the disease allele in CF families occurred also in normal individuals. The combined linkage studies, and the careful analysis of families in which recombination had occurred, indicated that two of the markers used (MET and D7S8) probably flanked the CF gene and hence defined the interval on chromosome 7 in which the gene must be. This result was of considerable significance for those groups trying to clone the gene itself.

LINKAGE DISEQUILIBRIUM AND THE ORIGIN OF CF MUTATIONS

Following the discovery of linkage much effort was put into the isolation of new DNA probes and the construction of a detailed physical map of the region. Many new markers were used to further the genetic analysis of CF families. The detection of recombination between a marker and the disease locus allows the latter to be more precisely localized, but as one gets closer to the disease gene the frequency of such recombinants decreases, and eventually this approach becomes uninformative.

Linkage disequilibrium is an alternative approach to genetic mapping over very small distances (usually less than 1% recombination). When a disease mutation first arises in an individual, it will be associated with a particular set of marker alleles (or haplotype) on the chromosome where the mutation occurred. For very closely linked markers, the chance of recombination is very small, so that the descendants of this original mutated chromosome will also tend to have the same marker haplotype. Therefore, there will be a difference in the distribution or frequency of the haplotype between normal and disease-carrying chromosomes. Such a difference is called linkage disequilibrium. Over a very long period of time, enough recombination will occur to disrupt the original relationship, and there will then be no significant difference between the haplotype frequency in normal and disease chromosomes. This is referred to as genetic (or linkage) equilibrium.

In a contemporary population, one may be able to observe linkage disequilibrium between a disease gene and closely linked markers, as long as neither the gene nor the markers are subject to subsequent new mutations at a significant frequency. In the case of a disease with an appreciable rate of new mutations, such as Duchenne muscular dystrophy, any linkage disequilibrium will soon disappear because new mutations giving rise to the disease can occur on chromosomes with any marker haplotype, and hence there will be no overall difference in the haplotype distribution between normal and disease chromosomes.

The large amount of linkage data collected by various laboratories all over Europe and the USA made CF an ideal candidate for linkage disequilibrium analysis. Four RFLPs were found to be in disequilibrium with CF, and of the 16 possible haplotypes two together accounted for 85% of CF chromosomes; however, the combined frequency of these two haplotypes in normal chromosomes was only 16%. This result suggested a common mutational origin for the majority of cases of CF in the European population. Later studies have confirmed this result. This demonstrates the power of linked markers, that are not themselves anything to do with the disease gene, in studying the origin of disease mutations.

TOWARD THE ISOLATION OF THE CF GENE

In addition to the purely genetic studies described above, there was of course much intense activity directed to the closing of the gap between the closest linked markers, and the identification of the disease gene itself. Most of the developing arsenal of new molecular genetic approaches were being applied. These included cloning as much of the region as possible using libraries of chromosome 7 derived by flow sorting and the use of somatic cell hybrids, and then mapping these new sequences using both linkage and physical methods. The technique of chromosome jumping was applied to extend the cloning and mapping in a directed manner, and the construction of long-range restriction maps of the CF region, using pulsed-field gel electrophoresis, provided a physical basis for the interpretation of the linkage disequilibrium results described above.

The laboratory at St Mary's Hospital in London had constructed a somatic cell hybrid containing the CF region of chromosome 7 as its only human component, by taking advantage of the fact that the MET oncogene, which is closely linked to CF, could be used as a marker selectable in cell culture. They made a specialized DNA library from this hybrid, using for the cloning step the restriction enzyme XmaIII, which only cleaves its recognition site (CCCGGG) when the CG doublet is unmethylated. Cleavage usually only occurs within HTF islands, the short, CG-rich, under-methylated regions associated with the 5' ends of genes. Thus the library that was made provided a means of accessing genes directly. A gene mapping within the region known to contain CF was thus isolated (Estivill *et al.*, 1987), and was considered to be a candidate for CF itself. However, subsequent extensive linkage studies of this gene (designated IRP) in CF families detected recombinations between the two loci, effectively ruling out IRP as being the CF gene. The recombination rate between CF and IRP was estimated at about 0.1 cM, linkage disequilibrium was detected, and the region remaining as the site of the locus was narrowed further.

ISOLATION OF THE CF GENE

The cloning of the CF gene was finally achieved in 1989 by a consortium of laboratories from Toronto, Canada, and Michigan, USA (Rommens *et al.*, 1989; Riordan *et al.*, 1989; Kerem *et al.*, 1989). Firstly, using a combination of chromosome walking and jumping, a region of 500 kb of DNA was cloned. This involved the isolation and characterization of hundreds of lambda and cosmid clones. This detailed physical map was used to position the various linked markers, including IRP, and by using the results of previous genetic studies the most likely location of CF was predicted. Potential coding sequences were obtained from this region by making use of HTF islands and by screening the cloned DNAs with Southern blots of DNA from a variety of species to identify sequences conserved through evolution and therefore probably having some functional significance. The characterization of the candidate genes so obtained involved hybridization to mRNA of tissues affected in CF, isolation of the corresponding cDNA clones and identification of open reading frames, and comparison of the sequence between normal and CF chromosomes.

The probes that eventually were shown to identify the CF gene were first considered interesting because of their high degree of cross-hybridization with animal DNA. They did not, however, detect any mRNA transcripts. Sequencing the DNA of these clones revealed an HTF island and some short regions of coding sequence. Various cDNA libraries were screened and eventually a clone was obtained, from a library made from normal cultured sweat gland cells. By screening further cDNA libraries, overlapping clones were obtained and eventually the whole gene was assembled. The gene consisted of at least 24 exons (there are now known to be 27) and covered about 250 kb of chromosomal DNA, most of which was introns. The gene was not detectably deleted or rearranged in a series of CF patients, suggesting that most cases of CF would be due to point mutations or very small deletions. This is in marked contrast to the situation for Duchenne/Becker muscular dystrophy, where the high frequency of large deletions played such a crucial role in the identification of the gene.

THE CF GENE MUTATION

Further characterization of the presumed CF gene was done by extensive cDNA cloning and sequencing (Riordan et al., 1989). The mRNA detected using the cDNA clones was about 6.5 kb in length, and the DNA sequence predicted a protein product of 1480 amino acids. Transcripts were detected in a variety of tissues, including pancreas, nasal polyps, lung, colon, sweat gland, placenta, liver, kidney and parotid gland, but not in brain or adrenal gland, nor in cultured fibroblasts or lymphoblasts. This distribution reflects the spectrum of tissues affected in CF, but this information alone was not enough to establish the candidate as the true CF gene. Since there was no apparent difference in the size or amount of the corresponding mRNA between CF and normal tissues, it was assumed that the CF mutation might be merely a single base change. Clones corresponding to the CF candidate gene were isolated from a cDNA library made from a CF patient, and shown by sequencing to have a deletion of three bases. According to the predicted protein sequence, the effect of this mutation would be the loss of amino acid 508, a phenylalanine residue. The mutation is called ΔF508.

In order to validate this finding, an assay was devised to test for the mutation in CF and normal chromosomes. Two oligonucleotide probes—one corresponding to the region of the gene sequence carrying the 3 bp deletion, the other representing its normal counterpart—were hybridized to polymerase chain reaction (PCR)-amplified samples of DNA from patients and controls. Of the 200 control chromosomes, all hybridized to the normal sequence oligo, and none to the mutated sequence oligo. Of the 200 CF chromosomes, about 70% hybridized to the mutated sequence oligo but not the normal sequence oligo. These results established beyond reasonable doubt that the mutation in the candidate gene was the cause of CF in 70% of cases, since the mutation was not found in normal samples, and that the gene that had been isolated was indeed the CF gene (Kerem et al., 1989). The mutation is now routinely detected using PCR amplification and gel electrophoresis as shown in Fig. 3.1.

The major clinical categorization of CF is into pancreatic-sufficient (PS) and pancreatic-insufficient (PI) subgroups, depending on the degree of impairment of the pancreas. Almost all of the patients who were homozygous for the

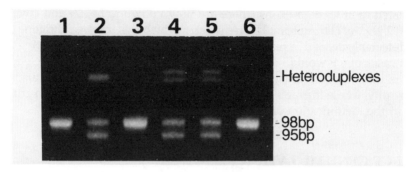

Fig. 3.1. Detection of the 3 bp deletions ΔF508 and ΔI507 within exon 10 of the CFTR gene. PCR products generated using primers C$_{16}$B and C$_{16}$D (Riordan *et al.*, 1989) are electrophoresed on a 10% polyacrylamide gel; a normal gene generates a 98 bp fragment, and a ΔF508 or ΔI507 mutated gene generates a 95 bp product. Furthermore in heterozygotes, a heteroduplex is formed between the normal and deleted fragments, which runs as additional bands of slower mobility. This heteroduplex provides a convenient means of distinguishing between the adjacent 3 bp mutations, since ΔF508 is seen as two discrete bands whereas ΔI507 is seen as a single band. Lanes 1, 3 and 6 are normal homozygotes, lane 2 is a ΔI507 heterozygote, and lanes 4 and 5 are ΔF508 heterozygotes

ΔF508 mutation were of the PI type. Conversely, the PS patients were generally compound heterozygotes, having a ΔF508 mutation on one chromosome and another, at that time undefined, CF mutation on their other chromosome. Thus the ΔF508 mutation may be regarded as a 'severe' CF allele with regard to pancreatic status.

The existence of one major CF mutation had already been predicted from the linkage disequilibrium studies, and it was found that ΔF508 was associated almost exclusively with a single haplotype of closely linked RFLPs. The remaining 30% of CF chromosomes were associated with a number of different haplotypes, suggesting that there could be at least seven additional mutations (Kerem *et al.*, 1989). At the time of writing, over 150 different mutations had been discovered (Tsui, 1992), and their distribution in the CF gene is shown in Fig. 3.2.

CFTR: THE CF GENE PRODUCT

IS CFTR A CHLORIDE CHANNEL?

The aim of many of the laboratories working on CF is to understand the role of the gene product in the disease process, and ultimately to design an effective therapy. The isolation of the gene and prediction of the structure of its protein product (Riordan *et al.*, 1989) may be seen as the first stage in this process. The protein consists of 1480 amino acids and has a molecular mass of 168 138 Da.

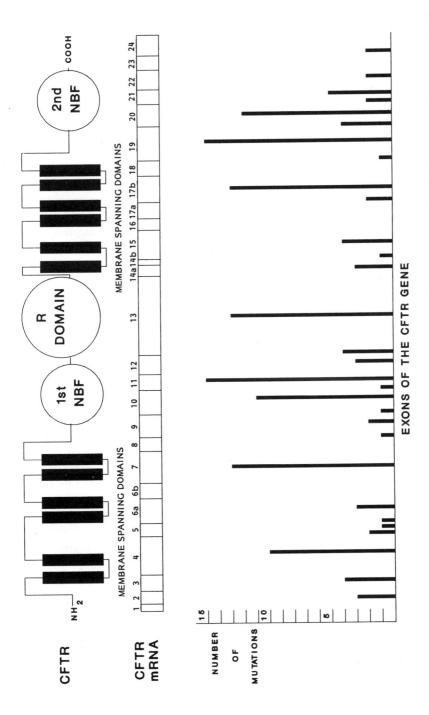

Fig. 3.2. The domains of the CFTR protein, the corresponding exons in the CFTR gene, and the distribution of disease-associated mutations. Abbreviations: NBF, nucleotide binding fold; R, regulatory; CFTR, cystic fibrosis transmembrane conductance regulator

Inspection of the sequence revealed two repeated motifs, each of which contains regions capable of spanning a membrane, and sequences resembling consensus nucleotide-binding folds. These characteristics are remarkably similar to those of the mammalian multidrug resistance P-glycoprotein and a number of other membrane-associated proteins, suggesting that the predicted CF gene product is probably a member of a membrane protein superfamily. The two motifs are joined by a regulatory domain, which has many potential sites for phosphorylation by protein kinases. These characteristics suggest that the CF protein is probably situated in the cell membrane, and may be involved in ion transport. In recognition of this the protein was named CFTR (cystic fibrosis transmembrane conductance regulator).

The phenylalanine residue deleted by the ΔF508 mutation was predicted to lie within the first putative ATP binding domain, and this suggested a disease mechanism: the loss of phenylalanine may prevent proper binding of ATP, or the conformational change required for normal CFTR activity, resulting in abnormal regulation. Comparison of the sequence with other proteins could not provide a definite role for CFTR in ion conductance. It was possible that CFTR is the ion channel itself, yet it was obviously not the same as any of the known chloride channel proteins, although some structural similarities with known ion channel proteins such as the brain sodium channel and the GABA receptor chloride channel were seen. Alternatively, it was suggested that CFTR might be associated with the actual ion channel in such a way as to regulate its activity (Riordan et al., 1989).

Hyde et al. (1990) developed a model for the common nucleotide binding domain in the ATP-binding cassette (ABC) superfamily of transport systems (which includes prokaryotic transporters, multidrug resistance proteins, and CFTR). Since F508 lies in a region believed responsible for coupling ATP-dependent conformational changes to the transport process, but not for binding ATP itself, they argued against the suggestion that ΔF508 CFTR is defective in binding or hydrolysing ATP. Other studies suggested that F508 was localized to a β-strand within the nucleotide binding domain and deletion of this residue would induce a significant structural change in the β-strand and altered nucleotide binding.

The isolation of the CF gene has made it possible to design experiments to test its function. Recombinant vaccinia virus vectors containing either the normal CFTR coding sequence or the ΔF508 mutated version were introduced into cultured CF airway epithelial cells, in which the normal or mutated CFTR was then expressed (Rich et al., 1990). By means of a fluorescent dye assay and single-cell patch-clamp techniques the effects of these constructs on ion transport were studied. It was found that the construct expressing the normal CFTR protein, but not that with the mutant version, could restore the chloride transport to at least normal levels. The effect was further stimulated by cAMP, consistent with previous studies on the effect of this nucleotide in vivo. Thus, these results confirmed the prediction that defects in CFTR are the cause of CF, via their effects on ion (particularly chloride) transport, but could not distinguish whether CFTR is the ion channel itself or is a regulator of such channels.

Subsequently it was shown that introduction and expression of the CF gene in non-epithelial cells (which are not generally affected in CF and presumably do not normally express the gene) endowed them with a plasma membrane chloride conductance that could be switched on by cAMP. Single-channel measurements made on cell-attached patches demonstrated that virtually all CFTR-expressing Sf9 insect cells displayed an 8.4 pS linear chloride channel on activation by cAMP. The simplest interpretation is that the protein itself is a cAMP-activated chloride channel. The alternative interpretation, that CFTR directly or indirectly regulates chloride channels, required that these cells had endogeneous cryptic chloride channels that were stimulated by cAMP only in the presence of CFTR. In CFTR-expressing Chinese hamster ovary cells chloride channels are produced that are regulated by analogues of cAMP, and by protein kinases A and C and alkaline phosphatase. This provides further evidence that CFTR is a chloride channel that can be activated by phosphorylation and inactivated by dephosphorylation, and reveals a synergism between converging kinase regulatory pathways.

Anderson et al. (1991a), and Rich et al. (1991) provided compelling evidence that the protein encoded by the CF gene is a regulated chloride channel. They introduced specific amino acid changes that altered the ion transport properties of CFTR, and effectively excluded the possibility that CFTR activates an endogenous channel by acting as the transporter of some regulatory molecule. In the first study positively charged amino acids in the transmembrane domains were separately replaced with negatively charged residues. This altered the ion selectivity of the channel in favour of I⁻ rather than Cl⁻. The simplest interpretation was that these charged residues contribute to a transmembrane 'pore' and play a direct part in determining ion selectivity. In the second study deletion of the regulatory (R) domain of CFTR was found to leave the channel permanently open, implicating this domain in the opening and closing of the channel in response to cAMP-activated protein kinases. The observation that a mutation in one of the ATP-binding domains was suppressed by deletion of the R domain suggests that the ATP-binding domains may also have a regulatory role: perhaps the R domain is moved into and out of the channel in an ATP-dependent fashion. In total, these data were convincing and supported the notion that CFTR is itself a regulated chloride channel or, at the very least, a component of a multisubunit channel.

The function of CFTR was definitely resolved by Bear et al. (1992), who purified the protein to homogeneity, reconstituted it into synthetic phospholipid vesicles, and fused these with planar lipid bilayers. This preparation exhibited regulated chloride channel activity, providing further evidence that the protein itself is the channel. This activity exhibited the basic biophysical and regulatory properties of the type of Cl⁻ channel found exclusively in normal CFTR-expressing cell types. This work also showed that the protein can be removed from the membrane, manipulated extensively, and returned to a functional state. These are important prerequisites to the potential use of purified CFTR for studies of its molecular structure and mechanisms of action, and in protein therapy.

WHAT ARE THE ROLES OF THE REGULATORY AND NUCLEOTIDE BINDING DOMAINS?

Phosphorylation of the regulatory domain of CFTR appears to regulate the Cl⁻ channel activity, since (1) addition of the catalytic subunit of protein kinase A (PKA) to the cytosolic surface of excised, cell-free patches of membrane activates the CFTR Cl⁻ channel; (2) four serines in the R domain are substrates for cAMP-dependent PKA *in vitro* and are phosphorylated *in vivo* when cellular levels of cAMP increase, and mutation of these serines to alanine prevents cAMP-dependent activation of CFTR Cl⁻ channels; (3) expression of CFTR from which most of the R domain has been deleted generates Cl⁻ channels that are active in the absence of cAMP.

Anderson *et al.* (1991b) considered the possibility that the CFTR nucleotide binding domains (NBDs) may regulate the channel. They argued convincingly that ATP is required during phosphorylation of the CFTR Cl⁻ channel by PKA and established that channel opening requires ATP hydrolysis in a second distinct mechanism independent of PKA (Fig. 3.3). Studies of the CFTR mutant K1250M (in which mutation of a lysine residue is thought to prevent ATP hydrolysis by NBD2) showed that ATP reversibly opened phosphorylated CFTR-K1250M channels. This result indirectly suggests that the influence of hydrolysable nucleoside triphosphates on NBD1 is sufficient to open CFTR. Because mutation of the equivalent residue in NBD1 generates an improperly processed protein, and accordingly no channel activity, Anderson *et al.* (1991b) could not ascertain whether NBD2 might be sufficient for ATP-dependent channel activation. Although Rich *et al.* (1991) had shown that the deletion of the R domain from CFTR generated Cl⁻ channels that were active without an increase in cAMP, these experiments were always performed in the presence of ATP. Anderson *et al.* (1991b) showed that ATP hydrolysis was also required to increase conductance in CFTR mutants lacking most of the R domain. It was proposed that ATP most likely regulates CFTR via one or both NBDs—an attractive conclusion since it provided a functional role for the NBDs of CFTR.

DEFECTIVE PROCESSING OF MUTANT CFTR

Recombinant vectors containing normal and various mutated versions of the CFTR gene were used to study the properties of the gene product in cultured

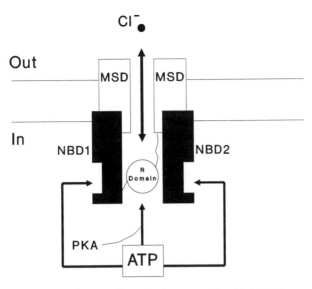

Fig. 3.3. Hypothetical model (taken from Anderson *et al.*, 1991b) of the mechanism of action of CFTR. Abbreviations: Cl⁻, chloride ions; MSD, membrane spanning domains; NBD, nucleotide binding domain; PKA, protein kinase A; R, regulatory. Double-headed arrow indicates chloride channel

COS-7 cells (Cheng *et al.*, 1990). The normal gene product is post-translationally modified by the addition of carbohydrate residues (glycosylation). However, if the introduced CFTR gene has ΔF508, the glycosylation of the gene product is inhibited. Not all the mutations tested showed this behaviour. The protein that remains unmodified is degraded in the endoplasmic reticulum or lysosomes. Thus it must be possible for the cell to recognize the mutant protein at this early stage; it is not clear how such a 'quality control' mechanism might operate. The presence of the precursor and mature forms of CFTR might even be used as an indicator of disease status, and hence as a diagnostic test. This might circumvent some of the difficulties encountered when the mutation in the subject is not one of those already characterized.

Further evidence for defective processing of mutant CFTR came from studies of normal airway epithelial cells, in which CFTR was found largely at the apical surface, whereas in cells derived from ΔF508 patients it was inside the cell. Furthermore, by the use of a panel of monoclonal antibodies to different domains of CFTR, it was shown that CFTR was especially prominent in the apical membrane of the reabsorptive sweat duct of normal biopsies, but was absent from the apical membrane of ΔF508 homozygote samples and reduced in those of heterozygous carriers. The parallel with the *in vitro* findings in COS-7 cells was sustained in studies of a CF patient containing the mutation G551D in one allele. In COS cells, G551D CFTR was processed correctly, and in contrast to ΔF508 CFTR, this protein was detected on the apical membrane *in vivo*. In addition, a number of CF homozygote sweat glands showed evidence for dense, 'granular' staining within some cells, including the sweat coil, again indicative of a processing defect.

The conclusions that the ΔF508 mutant protein is not processed correctly and, as a result, is not delivered to the plasma membrane, are consistent with earlier functional studies which failed to detect cAMP-stimulated Cl⁻ channels in cells expressing ΔF508 CFTR. In contrast however, chloride channel activity was detected when ΔF508 was expressed in *Xenopus* oocytes, Vero cells, or Sf9 insect cells. Furthermore, these studies indicated that cells expressing mutant CFTR became responsive to stimulation, thus offering hope for therapies based on improving the function of the mutant CFTR.

This raised the possibility that the apparent accumulation of mutant CFTR in cultured COS cells was an artefact of the expression system. Recent work resolves this issue. Oocytes and Sf9 cells are typically maintained at lower temperatures than mammalian cells, and processing of nascent proteins can be sensitive to temperature. It was found that the processing of ΔF508 CFTR reverts towards that of wild-type as the incubation temperature is reduced, and when the processing defect is corrected cAMP-regulated Cl⁻ channels appear in the plasma membrane. These results reconcile previously incongruent observations and suggest that ΔF508 CFTR is a temperature-sensitive mutant. In the Vero cells (grown at 37°C), ΔF508 CFTR was produced using a very high-level expression system, and the channel function observed could have represented a small fraction of the mutant protein that escaped the cellular quality control mechanism to reach the plasma membrane.

Therefore, current findings strongly suggest that, in ΔF508 homozygous CF patients, the majority of the CFTR cannot reach the surface, thus potential pharmacological solutions to CF may first need to solve the problem of redirecting the processing of ΔF508 CFTR.

OTHER FUNCTIONS OF CFTR

Epithelial cells contain many apparently independent ion channels. Considerable effort has been invested in identifying and characterizing the specific ion channel(s) associated with CF, with a view to pharmacological intervention. The so-called 'outwardly rectifying chloride channel' (ORCC) was originally identified as a strong candidate for the primary defect in CF (Schoumacher *et al.*, 1987). Several groups reported that this channel could not be activated by PKA or protein kinase C in membranes from CF cells (in contrast to membranes from non-CF cells). As the ORCC could be activated in CF membranes by 'non-physiological' procedures, for example by large depolarizing voltage pulses, it was concluded that the channel itself was present in CF cells but that its regulation was defective.

When the CF gene was isolated in 1989, the possibility was raised that

CFTR might itself be the ORCC, as it had all the characteristics of a membrane protein. Subsequently CFTR has been shown to be a chloride channel, but the properties of the CFTR channel were shown to be entirely different from those of the ORCC. Expression of CFTR in various cell lines is not correlated with outwardly rectifying, depolarization-induced Cl⁻ channels. Expression of recombinant CFTR using adeno-associated virus vectors in CF bronchial epithelial cells corrects defective Cl⁻ secretion, induces the appearance of small, linear conductance Cl⁻ channels, and restores PKA activation of outwardly rectifying Cl⁻ channels. In other words, CFTR and ORCC are distinct chloride channels, but CFTR can regulate ORCC. These studies provide direct evidence that CFTR can influence the activity of another ion channel; thus it is perhaps easier to understand how loss of CFTR, a small conductance chloride channel, can lead to the considerable changes in transepithelial ion conductances that have been measured from membranes in CF cells. It is difficult to envisage possible pharmacological intervention, since in most cases of CF the mutant protein is trapped in the endoplasmic reticulum. But if the symptoms of CF result from the failure to regulate other epithelial ion channels rather than the absence of the small conductance chloride channel *per se*, pharmacological intervention to restore regulation to these other channels becomes a possibility.

The CFTR gene is also expressed in B lymphoblasts, which suggests that CFTR has functional effects in lymphocytes that could be compromised by CFTR mutations. Transfection of CF lymphocytes with wild-type CFTR cDNA restores normal Cl⁻ conductance. This conductance is regulated in a manner coordinated with the cell cycle. However, the electrical characteristics of Cl⁻ conductance associated with CFTR expression in lymphocytes differs from that in other cells, which may result from lymphoid-specific regulatory influences on the properties of the conductance pathway of CFTR itself, or from the regulation, by CFTR, of a different Cl⁻ channel from that activated when CFTR is expressed in cells other than lymphocytes.

POPULATION ANALYSIS OF THE MAJOR CF MUTATION

In a collaborative European study, 4871 CF and 3539 normal chromosomes were characterized for the haplotypes defined by the two extragenic RFLPs XV2c and KM19 at the D7S23 locus. The association between one of the haplotypes (B) and the most frequent CF mutation, ΔF508, suggested for the latter a single origin and subsequent diffusion along a southeast–northwest gradient across Europe. The data showed considerable variation in the relative frequency of ΔF508, from 27% in Turkey to 88% in Denmark. Lower frequencies were observed in the southern populations, in agreement with previous reports suggesting that heterogeneity of mutations might be greater in southern European than northern European populations.

The relatively low frequency of ΔF508 in countries such as Italy, where the incidence of CF is quite high, suggests that genetic drift is not responsible for the high frequency of CF. Whereas it is possible for a single mutation on a chromosome of a rare haplotype to spread by chance alone, it is unlikely that this would take place more than once, giving rise to a population with a high frequency of CF but a low frequency of ΔF508. The idea of a selective advantage of CF heterozygotes remains the most plausible explanation for the high frequency of the CF gene. Heterozygotes may have an increased resistance to Cl⁻-secreting diarrhoeas especially in early infancy. Defective acidification in CF cells, caused by diminished Cl⁻ conductance, may give rise to a number of changes resulting in greater resistance to cholera, thereby possibly providing an alternative mechanism of heterozygote advantage. Another type of selective force, suggested by the strong residual association observed between non-ΔF508 mutations and haplotype B, might be due to a gene with a selective advantage located very close to CFTR and in strong association with this haplotype.

SCREENING FOR CF GENE CARRIERS

The discovery of the CF gene and the ability to detect easily carriers with the ΔF508 deletion, led, perhaps inevitably, to calls for population screening. The single major benefit of CF carrier screening is the option to make more informed reproductive decisions. However, several important issues need to be considered.

There are marked variations in the frequency of ΔF508 and some of the rarer CF mutations in different geographical locations and between different ethnic groups. For example, ΔF508 accounts for approximately 70% of all CF chromosomes in northern Europe but only 30% in southern Europe. Conversely, G542X, a nonsense mutation in exon 11, accounts for approximately 2.5% of the Scottish and Welsh CF chromosomes but 7.9% of those in the Spanish population. Hence, carrier risk estimates developed for one population may not be extrapolated to another. In these instances counselling becomes problematic.

Since many of the CF mutations are rare and many remain to date unidentified, it is not feasible to screen for all of them. However, 19 mutations account for more than 98% of CF chromosomes in a Celtic population from Brittany, and so population screening on a large scale in northwestern European populations might be possible. However, in general it will be difficult to detect more than 90% of mutant CFTR alleles except in ethnically and geographically discrete populations where CF is a result of founder effect. In the Welsh population about 69% of the couples at risk can be identified using a combined assay for four of the more common mutations. Some advocate that it would be unethical to neglect the possibility of informing those couples with a 1 : 4 risk of having a CF child that antenatal diagnosis is available to them. An early start on CF carrier screening (provided there is a back-up infrastructure of genetic counselling and support) would assist with public education, encouraging people to opt for carrier testing before pregnancy and widening their options for the future. However, others believe that until assays are available for all the CF mutations, such testing would have a significant false-negative rate, potentially resulting in enormous counselling and medical liability problems.

The American Society of Human Genetics adopted a position on CF population screening which states that routine carrier screening is not the standard of care. The Society acknowledged that testing should be offered to couples who have a close relative with CF and that pilot studies should be completed prior to the initiation of large-scale screening. Pilot studies in the UK seek to evaluate, among other things, the uptake of CF carrier testing offered through primary health care services. Advantages of this type of programme are that the information about CF, counselling, and the test itself are separate from immediate decisions regarding reproduction; this approach allows time for reflection. Results of tests on 1000 patients of reproductive age indicated that most people opportunistically offered a CF test would accept, and would use knowledge of carrier state in future reproductive decisions. Varying degrees of anxiety were found to be associated initially with a positive result, but most of this was allayed by genetic counselling, and no long-term psychological consequences in carriers were observed. It should be noted, however, that in using this approach of screening through primary health care services, only 35% of at-risk couples would be detected.

A second proposal is to offer screening to women who are attending antenatal clinics, which should identify 52% of the couples at potential risk. This approach has the advantage of higher uptake rates, and the information is imparted in the direct context in which it is to be used, thereby minimizing the risk of misinterpretation, forgetfulness and stigma. However, the testing and subsequent decisions must be quick to allow termination in the first or early second trimester. Pregnancy is a time of stress, and this could make the couple act hastily and ultimately regret their decision. It was found that there was a significant increase in stress at the time of the test result among women identified as carriers, which disappeared if their partners tested normal.

However, because of the practical problems referred to earlier, because standards of health care continue to improve, with great benefits to the quality of life available to a CF patient, and because of likely future advances in therapy, many professionals feel that general population screening is still premature.

GENOTYPE/PHENOTYPE CORRELATIONS

ΔF508

It was originally proposed that ΔF508 is a severe allele (with regard to pancreatic status), that pancreatic insufficiency (PI) is due to the presence of two severe alleles and that a pancreatic-sufficient (PS) patient carries either a single severe allele or two mild alleles (Kerem *et al.*, 1989). Some have claimed that there does not appear to be a significant correlation between ΔF508 and severity of the lung disease; other factors, genetic or non-genetic, may affect the severity of disease symptoms, but these factors do not necessarily correlate with specific CF mutations. However, others found a significant decline in lung function, earlier onset of symptoms, greater need for pancreatic enzyme substitution and increased risk of chronic *Pseudomonas aeruginosa* infection in ΔF508 homozygotes relative to heterozygotes.

NONSENSE MUTATIONS

Nonsense mutations cause premature termination of mRNA translation and have been associated with abnormally low levels of mutant mRNA. Usually an unstable truncated protein results, frequently giving rise to severe illness. However, recent reports have been presented in discordance with these observations. Two patients with nonsense mutations in each CF gene (G542X/S1255X and R553X/W1316X) showed severe pancreatic involvement but mild pulmonary disease, and children homozygous for G542X are only mildly affected. One patient homozygous for the R553X nonsense mutation was moderately severely affected, but another such had only mild pulmonary disease. Nine CF patients homozygous for the R1162X stop mutation had mild or moderate lung disease, but patients homozygous for W1282X are severely affected.

Further studies should clarify whether alternative RNA splicing, as has been demonstrated for exon 9 of normal CFTR, occurs to avoid some of the stop codons, or whether truncated proteins are produced which are either subsequently degraded or integrated into the cell membrane. It may be that major truncations make no stable CFTR, whereas proteins harbouring missense mutations might retain partial activity while trapped at incorrect cellular locations, causing a more general dysfunction than complete absence of protein.

THE MISSENSE MUTATION G551D

Compound heterozygotes for G551D/ΔF508 show less meconium ileus and a trend toward later age at diagnosis of pancreatic insufficiency than ΔF508 homozygotes. No significant difference was found for any other clinical parameter. The results suggested that the CF genotype can be a predictor of pancreatic and intestinal phenotype. It is difficult to explain the apparent lack of a major phenotypic difference between ΔF508 and G551D in light of the recent reports of ΔF508 CFTR being trapped in the endoplasmic reticulum, but G551D CFTR reaching the cell membrane.

OTHER MUTATIONS

A panel of 538 CF patients with well-characterized pancreatic function was screened for a series of known mutations. The results showed that each genotype was associated only with PI or only with PS, but not with both. The result is thus consistent with the hypothesis that PI and PS are predisposed by the CF genotype; the PS phenotype occurs in patients who have one or two mild CFTR mutations, whereas the PI phenotype occurs in patients with two severe alleles. In this context, all the stop codon, splice junction and frameshift mutations are classed as severe, whereas the missense mutations caused by single nucleotide changes could be severe or mild. The data strongly support the hypothesis that mild mutant alleles confer a higher residual CFTR activity in the pancreas than do severe mutant alleles. Furthermore, skin biopsies from two CF patients with mild disease (pancreatic sufficient) showed apical localization of CFTR.

THE MOUSE HOMOLOGUE OF THE CFTR GENE AND MOUSE MODELS

The deduced protein sequence of the mouse homologue of CFTR is 78% identical to the human protein, with higher conservation in the transmembrane and nucleotide-binding domains. The sites of known CF missense mutations are conserved between man and mouse, and many of the mutations in exons 10 and 11 of the CFTR gene occur at highly conserved codons in five different mammalian species, suggesting that these sites play an essential role in the functioning of the CFTR protein. The predicted mouse protein has a phenylalanine residue corresponding to that deleted in ΔF508, and a mouse model with this mutation is currently being developed.

An animal model for CF would be very useful for testing new treatment modalities and for further investigations of the pathology of the disease. Gene targeting is a method whereby all or part of the host cell gene is replaced by a modified version, constructed *in vitro* and introduced into embryonic stem cells by microinjection or electroporation (Fig. 3.4). The modified cells are then introduced into a blastocyst which is implanted in a surrogate mother mouse, resulting in the birth of chimeric offspring. Thus various CF mutations can be introduced into the mouse CFTR, and carrier mice are then interbred to produce offspring with different combinations of CF alleles.

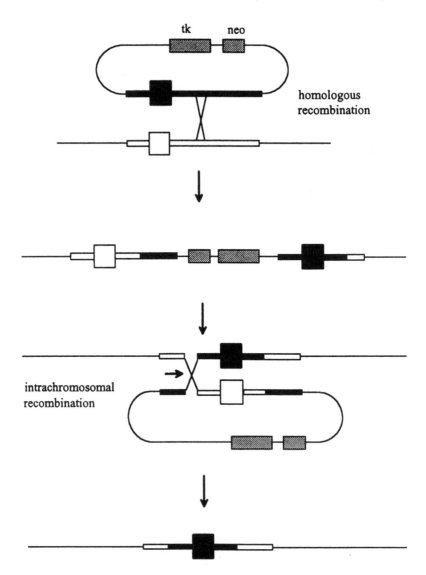

Fig. 3.4. Method for replacement of a chromosomal gene with a mutated version ('gene targeting'). Stage 1: a plasmid vector containing the mutated (black) version of the gene, positively and negatively selectable markers (neomycin resistance (neo) and thymidine kinase (tk) respectively) are introduced into a cell line by transfection or electroporation. Stage 2: integration of the plasmid into the chromosomal locus via homologous recombination, selected by neomycin resistance and checked by PCR for correct targeting. Stage 3: intrachromosomal recombination events that delete the plasmid sequences are selected by loss of the thymidine kinase marker. Depending on exactly where recombination occurs, a proportion of these will have replaced part of the wild-type gene (white box) with the mutated version (black box)

Mouse models for CF have been constructed by Snouwaert *et al.* (1992) and Dorin *et al.* (1992). In the first of these studies, two cell lines harbouring a premature stop codon in exon 10 of the CFTR gene were constructed using a 'replacement vector'. With this approach the vector used contains a mutated version of part of the CFTR gene, which replaces the corresponding part of the wild-type gene by homologous recombination. Mice carrying the mutation in their germ cells were identified and crossed to produce homozygous (−/−) animals. Initially there were fears that the (−/−) animals might have an embryonic lethal phenotype, since no naturally occurring CF mice have been identified. Fortunately, (−/−) mice were produced in a ratio expected from Mendelian inheritance; however, shortly after birth these animals began dying (few survived beyond 40 days). Intestinal obstruction, perforation and severe peritonitis was the cause of death in all the cases, bearing a striking similarity to meconium ileus in the human syndrome (which occurs in 5–10% of CF newborns and can be fatal). Upon examination, distension of the crypts of Lieberkuhn by dense concretions throughout the intestinal tract were evident, and the respiratory and intestinal epithelia lacked the normal cAMP-stimulated apical chloride channel, confirming the similarity to the human phenotype at the cellular level.

There were, however, poor pathological similarities between the respiratory, reproductive, pancreatic and hepatobiliary systems of the mouse model and the human disease. In the lungs, no evidence for obstructive disease was observed, which may reflect anatomical differences: rodents essentially lack the components of the human airway primarily responsible for mucus production. In the reproductive tract there was no evidence for abnormality in the mouse, and one (−/−) male successfully reproduced; nor was there conclusive evidence for pancreatic or hepatobiliary disease (except for inflammation of the gall-bladder). However, the possibility remains that disease manifestations might have appeared in the lungs, liver and pancreas, had the animals survived longer.

The study of Dorin *et al.* (1992) used a different strategy—insertional inactivation—to target a mutation to the mouse CFTR gene. A vector was used that contained parts of intron 9 and exon 10. This was integrated into the mouse gene by homologous recombination, resulting in a mutant allele in which parts of intron 9 and exon 10 were duplicated, and the vector sequences were inserted. (This type of experiment can also be modified to yield a mutant allele specifically altered at a single site, if the incoming vector DNA contains the desired mutation and is subsequently allowed to remove itself from the host locus by intrachromosomal recombination, leaving only the mutated segment of CFTR behind; see Fig. 3.4.) Mice homozygous for the insertional inactivation showed some of the signs of human CF, including altered Cl⁻ ion transport and pathological changes to the colon, vas deferens, lungs and salivary ducts. However, these latter manifestations were very variable and there was no clinical disease in the mutant mice, and neither did they suffer from bowel obstructions leading to early death. The relative mildness of the phenotype might have been partly due to the production of very low levels of wild-type CFTR message by aberrant RNA splicing.

Although there might be approaches to ameliorating the disease phenotype in CF mice that would be beneficial in treatment of the human disease, the anatomical and physiological differences between the two species indicate caution. To address many of the questions in CF it may well be necessary to produce mice with the ΔF508 mutation and other defined mutations, as these may have phenotypes different from those described above. Furthermore, the identification of other mildly affected animals with longer survival times that allow the emergence of a pulmonary phenotype will be essential to answer questions of the efficacy of various types of treatment.

THERAPEUTIC STRATEGIES

THE USE OF THERAPEUTIC AGENTS

Defects in the CFTR gene cause excessive sodium absorption and abolish chloride secretion in pulmonary epithelial cells. This contributes to the dehydration of airway secretions which are consequently poorly cleared, resulting in recurrent pulmonary infections (the major cause of death in CF). Amiloride, a sodium channel blocker, prevents sodium reabsorption in the epithelia of CF patients, and when administered in an aerosol form improves mucociliary clearance and reduces the rate of decline in vital capacity.

Since agents that stimulate chloride secretion via cAMP-mediated pathways are ineffective in CF patients, scientists investigated agents that acted independent of cAMP metabolism. Extracellular nucleotide triphosphates, when applied to the apical surface of cultured human airway epithelia, acted as effective chloride secretagogues. Furthermore, through studying nasal transepithelial potential differences, which give a measure of the rates of sodium absorption and chloride secretion, it was shown that nucleotide triphosphates also stimulated chloride secretion *in vivo* (probably via P2 nucleotide receptors). When applied in conjunction with amiloride, ATP and UTP, at their maximal effective concentrations of 10^{-4} M, were more effective chloride secretagogues in the CF patients than in the normal subjects. Of the two nucleotides UTP would be favoured as a therapeutic agent, administered with a sodium channel blocker, since ATP is rapidly degraded by ectonucleosidases to products that induce bronchoconstriction when inhaled by subjects with asthma. Clinical trials are in progress.

A similar strategy has been investigated at Stanford University. It had been previously reported that Cl⁻ flux in CF cells, as in normal cells, can be activated by raising intracellular calcium. This calcium activation of chloride channels is mediated in normal and CF cells via a Ca^{2+}/calmodulin-dependent protein kinase (CaMK). CF airway cells retain CaMK regulation of Cl⁻ current even though they lack PKA activation, and this difference may be exploited to develop agents that specifically stimulate phosphorylation of the Cl⁻ channel regulator by CaMK in epithelial cells. The defect in CF may be circumvented by the activation of this alternate pathway for regulation of Cl⁻ conductance.

The most problematic of the symptoms of CF is the secretion into the lung airways of large amounts of viscous and obstructive mucus. As well as impairing lung function directly, these also provide a rich breeding ground for bacteria. Hubbard *et al.* (1992) described a novel and elegantly simple strategy designed to overcome this clinical problem. They used human deoxyribonuclease, produced in large quantities by genetic engineering and administered to the patients by aerosol spray, to break down the DNA component of the mucus secretion and hence reduce its viscosity. The initial results were most encouraging as they showed a significant improvement in lung function in patients so treated.

GENE THERAPY

Since recombinant DNA technology began to be applied to human disease, gene therapy has been proposed as a means of treatment. It is important to draw a distinction between somatic cell gene therapy, in which a proportion of the cells in a target tissue (normally a tissue affected by disease) are treated, and germ-line therapy, in which the aim is to correct the defect in essentially all the cells of the organism, by manipulation at the earliest possible stage. Germ-line modifications are of course heritable in subsequent generations, whereas somatic cell therapy only affects the individual actually treated. Because of the experimental nature of these procedures, and because of the implications of interference with the human gene pool, human germ-line manipulation is not an acceptable option (although it is used as a research tool in animal genetics). Effective gene therapy of somatic cells depends on a number of considerations. These include accessibility of the target tissue, nature of the genetic defect, availability of a vector system capable of introducing the therapeutic gene, and correct expression of the gene once introduced. Progress on all these fronts, and the nature of the disease itself, has made CF a particularly attractive candidate for this approach.

Previous gene therapy experiments were largely based on the use of modified retroviruses as the gene transfer vector, since they could integrate their genes into the genomes of the cells they infect, thus providing a means of introducing a wild-type version of the disease gene. Assuming that the gene is expressed appropriately, there is the potential for therapeutic production of the protein over a long period of time. However, with lung cells two problems exist with this approach. Firstly, retroviruses will integrate genes only into dividing cells, and since the majority of epithelial cells of the lung and other affected tissues have reached maturity and stopped dividing, the retrovirally transmitted foreign gene will not be acquired. Secondly, gene transfer with retroviruses usually involves removing target cells from the patient, incubating them with the vector, and subsequently returning them; however, lung cells are not sufficiently accessible to be treated via this approach.

At the National Heart, Lung and Blood Institute, a new strategy was developed based on an adenovirus vector, which infects lung epithelial cells naturally, and does not require host cell proliferation for the expression of its proteins (Rosenfeld *et al.*, 1992). Furthermore, there are no known associations of human malignancies with adenoviral infections, recombination is rare, the genome can be manipulated to accommodate foreign genes of up to 7.0–7.5 kb in length, and live adenovirus has been used safely as a human vaccine.

Rosenfeld's group used a replication deficient adenoviral vector (Ad-α1AT) containing an adenovirus major late promoter and a recombinant human α_1-antitrypsin gene to infect epithelial cells of the cotton rat respiratory tract *in vitro* and *in vivo*. Following *in vivo* intratracheal administration of Ad-α1AT, human α1AT mRNA was observed in the respiratory epithelium, and human α1AT was synthesized and secreted by lung tissue and was detected in the epithelial lining fluid for at least 1 week. These encouraging results suggested that similar recombinant vectors might be useful for *in vivo* experimental animal studies with the human CF gene. Furthermore, *in vitro* studies (Rich *et al.*, 1990) had previously demonstrated that the transfer of normal CFTR cDNA to CF epithelial cell lines corrects the CF defect, thus permitting the cells to secrete Cl⁻ in response to intracellular cAMP.

Subsequently a replication-deficient recombinant adenovirus (Ad-CFTR) containing a normal CFTR cDNA was used, to demonstrate *in vivo* transfer and expression of the human CFTR gene to the respiratory epithelium of the lungs of cotton rats. Following intratracheal administration of Ad-CFTR the presence of human CFTR mRNA transcripts was detected in bronchial epithelium by *in situ* hybridization analysis using a human CFTR cRNA probe. The human origin of the transcripts was confirmed by PCR, and northern analysis of lung RNA showed Ad-CFTR-directed human CFTR mRNA transcripts at up to 6 weeks after *in vivo* infection. Most importantly, immunohistochemical evaluation with an anti-human CFTR antibody demonstrated that human CFTR protein was present 11–14 days after *in vivo* infection with Ad-CFTR.

More recently, Hyde *et al.* (1993) demonstrated that in a transgenic CF mouse the ion transport defect could be corrected by delivery of a CFTR expression plasmid complexed with liposomes into the lungs. RNA analysis was used to show that the CFTR gene was expressed in the airway epithelium, and that not only the primary ion transport deficiency (cAMP-stimulated chloride transport) but also secondary alterations in sodium absorption could be corrected. The authors suggested that a similar strategy could be applied to human patients.

Although these studies demonstrate the feasibility of using artificial vector systems to transfer a normal CFTR cDNA to the lungs, several questions have to be addressed in the context of applying these therapies to individuals with CF:

(1) The threshold levels of expression and the specific epithelial cell targets necessary to reverse the disease process must be determined. The current working hypothesis is that gene therapy for CF will require only low-level expression of normal CFTR to correct the defective physiology in the airway epithelium.
(2) The consequences of excess expression of the normal CFTR gene in the airway epithelial cells are unknown, and the possible need for regulated expression will have to be addressed.
(3) The safety of the vector system is important. Although the Ad-CFTR vector was replication deficient, the possibility remains that the host cell could provide the missing function in *trans*, thus allowing viral replication. Alternatively, a wild-type adenovirus infection might complement or recombine with the Ad-CFTR vector, giving it the ability to replicate.
(4) In the case of the Ad-CFTR vector, is not known what proportion, if any, of the recombinant DNA is integrated into the genome of the target cells; thus the persistence of coexpression is unknown.

However, it is likely that many of these problems will be solved by experimental approaches, and that gene therapy may well become one of a number of possible treatments for CF. In Europe and the USA, clinical trials are now in progress with both the liposome-mediated expression system and the viral vector approach. Indeed, CF is a particularly suitable disease for the application of gene therapeutic approaches, due to its recessive mode of inheritance (whereby the disease state is due to loss of a particular function) and easy accessibility of the lung cells. The application of molecular genetic approaches to CF, leading to the isolation of the gene, characterization of the spectrum of mutations, elucidation of the function of the gene product, and now to novel strategies for therapy, is one of the most gratifyingly successful stories in medical genetics to date.

REFERENCES

Anderson MP, Gregory RJ et al. (1991a) Demonstration that CFTR is a chloride channel by alteration of its anion selectivity. *Science*, **253**, 202–205.
Anderson MP, Berger HA et al. (1991b) Nucleotide triphosphates are required to open the CFTR chloride channel. *Cell*, **67**, 775–784.
Bear CE, Li C et al. (1992) Purification and functional reconstitution of the cystic fibrosis transmembrane conductance regulator (CFTR). *Cell*, **68**, 809–818.

Cheng SM, Gregory RJ et al. (1990) Defective intracellular transport and processing of CFTR is the molecular basis of most cystic fibrosis. Cell, 63, 827–834.

Dorin JR, Dickinson P et al. (1992) Cystic fibrosis in the mouse by targeted insertional mutagenesis. Nature, 359, 211–215.

Eiberg H, Mohr J et al. (1985) Linkage relationships of paraoxonase (PON) with other markers: indication of PON-cystic fibrosis synteny. Clin. Genet., 28, 265–271.

Estivill X, Farrall M et al. (1987) A candidate for the cystic fibrosis locus isolated by selection for methylation-free islands. Nature, 326, 840–845.

Hubbard RC, McElvaney NG et al. (1992) A preliminary study of aerosolized recombinant human deoxyribonuclease I in the treatment of cystic fibrosis. N. Engl. J. Med., 326, 812–815.

Hyde SC, Emsley P et al. (1990) Structural model of ATP-binding proteins associated with cystic fibrosis, multidrug resistance and bacterial transport. Nature, 346, 362–365.

Hyde SC, Gill DR et al. (1993) Correction of the ion transport defect in cystic fibrosis transgenic mice by gene therapy. Nature, 362, 250–255.

Kerem B, Rommens JM et al. (1989) Identification of the cystic fibrosis gene: genetic analysis. Science, 245, 1073–1080.

Knowlton RG, Cohen-Haguenauer O et al. (1985) A polymorphic DNA marker linked to cystic fibrosis is located on chromosome 7. Nature, 318, 380–382.

Rich DP, Anderson MP et al. (1990) Expression of cystic fibrosis transmembrane conductance regulator corrects defective chloride channel regulation in cystic fibrosis airway epithelial cells. Nature, 347, 358–363. (See also Gregory RJ, Cheng SM et al., ibid., 382–386.)

Rich DP, Gregory RJ et al. (1991) Effect of deleting the R domain on CFTR-generated chloride channels. Science, 253, 205–207.

Riordan JR, Rommens JM et al. (1989) Identification of the cystic fibrosis gene: cloning and characterisation of complementary DNA. Science, 245, 1066–1073.

Rommens JM, Iannuzzi MC et al. (1989) Identification of the cystic fibrosis gene: chromosome walking and jumping. Science, 245, 1059–1065.

Rosenfeld MA, Yoshimura K et al. (1992) In vivo transfer of the human cystic fibrosis transmembrane conductance regulator gene to the airway epithelium. Cell, 68, 143–155.

Schoumacher RA, Shoemaker RL et al. (1987) Phosphorylation fails to activate chloride channels from cystic fibrosis airway cells. Nature, 330, 752–754.

Snouwaert JN, Brigman KK et al. (1992) An animal model for cystic fibrosis made by gene targeting. Science, 257, 1083–1088.

Tsui LC, Buchwald M et al. (1985) Cystic fibrosis locus defined by a genetically linked polymorphic DNA marker. Science, 230, 1054–1057.

Tsui LC (1992) Mutations and sequence variations detected in the cystic fibrosis transmembrane conductance regulator (CFTR) gene. Hum. Mutat., 1, 197–203.

Wainwright BJ, Scambler PJ et al. (1985) Localisation of cystic fibrosis locus to human chromosome 7cen-q22. Nature, 318, 384–385.

White R, Woodward S et al. (1985) A closely linked marker for cystic fibrosis. Nature, 318, 382–384.

FURTHER READING

CLONING AND CHARACTERIZATION OF THE CFTR GENE

Anderson MP, Rich DP et al. (1991) Generation of cAMP-activated chloride currents by expression of CFTR. Science, 251, 679–682.

Beaudet A, Bowcock A *et al.* (1986) Linkage of cystic fibrosis to two tightly linked DNA markers: joint report from a collaborative study. *Am. J. Hum. Genet.*, **39**, 681–693.

Berger HB, Anderson MP *et al.* (1991) Identification and regulation of the CFTR-generated chloride channel. *J. Clin. Invest.*, **88**, 1422–1431.

Boucher R, Stutts M *et al.* (1986) Na⁺ transport in cystic fibrosis respiratory epithelia: abnormal basal rate and response to adenylate cyclase activation. *J. Clin. Invest.*, **78**, 1245–1252.

Cheng SH, Rich DP *et al.* (1991) Phosphorylation of the R domain by cAMP-dependent protein kinase regulates the CFTR chloride channel. *Cell*, **66**, 1027–1036.

Dalemans W, Barbry P *et al.* (1991) Altered chloride ion channel kinetics associated with the ΔF508 cystic fibrosis mutation. *Nature*, **354**, 526–528.

Denning GM, Ostedgaard LS *et al.* (1992) Abnormal localisation of cystic fibrosis transmembrane conductance regulator in primary cultures of cystic fibrosis airway epithelia. *J. Cell Biol.*, **118**, 551–559.

Denning GM, Anderson MP *et al.* (1992) Processing of mutant cystic fibrosis transmembrane conductance regulator is temperature sensitive. *Nature*, **358**, 761–764.

Drumm M, Smith C *et al.* (1988) Physical mapping of the cystic fibrosis region by pulsed-field gel electrophoresis. *Genomics*, **2**, 346–354.

Drumm ML, Pope HA *et al.* (1990) Correction of the cystic fibrosis defect in vitro by retrovirus-mediated gene transfer. *Cell*, **62**, 1227–1233.

Drumm ML, Wilkinson DJ *et al.* (1991) Chloride conductance expressed by ΔF508 and other mutant CFTRs in *Xenopus* oocytes. *Science*, **254**, 1797–1799.

Egan M, Flotte T *et al.* (1992) Defective regulation of outwardly rectifying Cl⁻ channels by protein kinase A corrected by insertion of CFTR. *Nature*, **358**, 581–584.

Farrall M, Wainwright BJ *et al.* (1986) Recombinations between IRP and cystic fibrosis. *Am. J. Hum. Genet.*, **43**, 471–475.

Frizzell R, Rechkemmer G *et al.* (1986) Altered regulation of airway epithelial cell chloride channels in cystic fibrosis. *Science*, **233**, 558–560.

Gregory RJ, Rich DP *et al.* (1991) Maturation and function of cystic fibrosis transmembrane conductance regulator variants bearing mutations in putative nucleotide-binding domains 1 and 2. *Mol. Cell. Biol.*, **11**, 3886–3893.

Hwang TC, Lu L *et al.* (1989) Cl⁻ channels in CF: lack of activation by protein kinase C and cAMP-dependent protein kinase. *Science*, **244**, 1351–1353.

Kartner N, Hanrahan JW *et al.* (1991) Expression of the cystic fibrosis gene in non-epithelial invertebrate cells produces a regulated anion conductance. *Cell*, **64**, 681–691.

Kartner N, Augustinas O *et al.* (1992) Mislocation of ΔF508 CFTR in cystic fibrosis sweat gland. *Nature Genet.*, **1**, 321–327.

Li M, McCann JD *et al.* (1988) Cyclic AMP-dependent protein kinase opens chloride channels in normal but not cystic fibrosis airway epithelium. *Nature*, **331**, 358–360.

Poustka A, Lehrach H *et al.* (1988) A long-range restriction map encompassing the cystic fibrosis locus and its closely linked genetic markers. *Genomics*, **2**, 337–345.

Rommens JM, Zengerling S *et al.* (1988) Identification and regional localisation of DNA markers on chromosome 7 for the cloning of the cystic fibrosis gene. *Am. J. Hum. Genet.*, **43**, 645–663.

Sarkadi B, Bauzon D *et al.* (1992) Biochemical characterisation of the cystic fibrosis transmembrane conductance regulator in normal and cystic fibrosis epithelial cells. *J. Biol. Chem.*, **267**, 2087–2095.

Schmiegelow K, Eiberg H *et al.* (1986) Linkage between the loci for cystic fibrosis and paraoxonase. *Clin. Genet.*, **29**, 374–377.

Schoumacher R, Ram J *et al.* (1990) A cystic fibrosis pancreatic adenocarcinoma cell line. *Proc. Natl Acad. Sci. USA*, **87**, 4012–4016.

Tabcharani JA, Chang X-B *et al.* (1991) Phosphorylation-regulated Cl⁻ channel in CHO cells stably expressing the cystic fibrosis gene. *Nature*, **352**, 628–631.

Thomas PJ, Shenbagamurthi P *et al.* (1991) Cystic fibrosis transmembrane conductance regulator: nucleotide binding to a synthetic peptide. *Science*, **251**, 555–557.

Ward CL, Krouse ME *et al.* (1991) Cystic fibrosis gene expression is not correlated with rectifying Cl⁻ channels. *Proc. Natl Acad. Sci. USA*, **88**, 5277–5281.

Welsh MJ (1986) An apical-membrane chloride channel in human tacheal epithelium. *Science*, **232**, 1648–1650.

Welsh MJ and Liedtke CM (1986) Chloride and potassium channels in cystic fibrosis airway epithelia. *Nature*, **322**, 467–470.

Willumsen N and Boucher R (1989) Activation of an apical Cl⁻ conductance by Ca^{2+} ionophores in cystic fibrosis airway epithelia. *Am. J. Physiol.*, **256**, C226–C233.

Zielenski J, Rozmahel R *et al.* (1991) Genomic DNA sequence of the cystic fibrosis transmembrane conductance regulator gene. *Genomics*, **10**, 214–228.

POPULATION ANALYSIS OF THE MAJOR MUTATION

CF Consortium (1990) Population analyis of the major mutation in cystic fibrosis. *Hum. Genet.*, **85**, 391–445.

Devoto M, De Benedetti L *et al.* (1989) Haplotypes in cystic fibrosis patients with or without pancreatic insufficiency from four European populations. *Genomics*, **5**, 894–898.

Estivill X, Scambler PJ *et al.* (1987) Patterns of polymorphism and linkage disequilibrium for cystic fibrosis. *Genomics*, **1**, 257–263.

Estivill X, Farrall M *et al.* (1988) Linkage disequilibrium between cystic fibrosis and linked DNA markers in Italian families: a collaborative study. *Am. J. Hum. Genet.*, **43**, 23–28.

Romeo G, Galietta JLV *et al.* (1989) Why is the CF gene so frequent? *Am. Hum. Genet.*, **84**, 1–5.

CARRIER SCREENING

Beaudet A (1990) Carrier screening for cystic fibrosis. *Am. J. Hum. Genet.*, **47**, 603–605.

Biesecker L, Bowles-Biesecker B *et al.* (1992) General population screening for cystic fibrosis is premature. *Am. J. Hum. Genet.*, **50**, 438–439.

Brock D (1990) Population screening for cystic fibrosis. *Am. J. Hum. Genet.*, **47**, 164–165.

Caskey CT, Kaback MM *et al.* (1990) The American Society of Human Genetics statement on cystic fibrosis screening. *Am. J. Hum. Genet.*, **46**, 393.

CF Consortium (1990) Population analysis of the major mutation in cystic fibrosis. *Hum. Genet.*, **85**, 391–445.

Cheadle J, Myring J *et al.* (1992) Mutation analysis of 184 cystic fibrosis families in Wales. *J. Med. Genet.*, **29**, 642–646.

Cutting GR, Curristin SM *et al.* (1992) Analysis of four diverse population groups indicates that a subset of cystic fibrosis mutations occur in common among caucasians. *Am. J. Hum. Genet.*, **50**, 1185–1194.

Ferec C, Audrezet MP *et al.* (1992) Detection of over 98% cystic fibrosis mutations in a celtic population. *Nature Genet.*, **1**, 188–191.

Ferrie RM, Schwarz MJ *et al.* (1992) Development, multiplexing, and applications of ARMS tests for common mutations in the CFTR gene. *Am. J. Hum. Genet.*, **51**, 251–262.

Gilbert F (1990) Is population screening for cystic fibrosis appropriate now? *Am. J. Hum. Genet.*, **46**, 394–395.
Lancet Editorial (1992) Screening for cystic fibrosis. *Lancet*, **340**, 209–210.
Mennie ME, Gilfillan A *et al.* (1992) Prenatal screening for cystic fibrosis. *Lancet*, **340**, 214–216.
Nunes V, Gasparini P, *et al.* (1991) Analysis of 14 cystic fibrosis mutations in five Southern European populations. *Hum. Genet.*, **87**, 737–738.
Shrimpton AE, McIntosh I *et al.* (1991) The incidence of different cystic fibrosis mutations in the Scottish population: effects on prenatal diagnosis and genetic counselling. *J. Med. Genet.*, **28**, 317–231.
Watson EK, Mayall ES *et al.* (1991) Screening for carriers of cystic fibrosis through primary health care services. *Br. Med. J.*, **303**, 504–507.
Watson EK, Mayall ES *et al.* (1992) Psychological and social consequences of community carrier screening programme for cystic fibrosis. *Lancet*, **340**, 217–220.
Wilfond BS and Fost N (1990) The cystic fibrosis gene: medical and social implications for heterozygote detection. *JAMA*, **263**, 2777–2783.

GENOTYPE/PHENOTYPE CORRELATIONS

Bal J, Stuhrmann M *et al.* (1991) A cystic fibrosis patient homozygous for the nonsense mutation R553X. *J. Med. Genet.*, **28**, 715–717.
Cheadle J, Al-Jader L *et al.* (1992) Mild pulmonary disease in a cystic fibrosis child homozygous for R553X. *J. Med. Genet.*, **29**, 597.
Cuppens H, Marynen P *et al.* (1990) A child, homozygous for a stop codon in exon 11, shows milder cystic fibrosis symptoms than her heterozygous nephew. *J. Med. Genet.*, **27**, 717–719.
Cutting GR, Kasch LM *et al.* (1990) Two patients with cystic fibrosis nonsense mutations in each cystic fibrosis gene and mild pulmonary disease. *N. Engl. J. Med.*, **323**, 1685–1689.
Gasparini P, Borgo G *et al.* (1992) Nine cystic fibrosis patients homozygous for the CFTR nonsense mutation R1162X have mild or moderate lung disease. *J. Med. Genet.*, **29**, 558–562.
Hamosh A, King TM *et al.* (1992) Cystic fibrosis patients bearing both the common missense mutation Gly > Asp at codon 551 and the ΔF508 mutation are clinically indistinguishable from ΔF508 homozygotes, except for decreased risk of meconium ileus. *Am. J. Hum. Genet.*, **51**, 245–250.
Johansen HK, Nir M *et al.* (1991) Severity of cystic fibrosis in patients homozygous and heterozygous for ΔF508 mutation. *Lancet*, **337**, 631–634.
Kristidis P, Bozon D *et al.* (1992) Genetic determination of exocrine pancreatic function in cystic fibrosis. *Am. J. Hum. Genet.*, **50**, 1178–1184.
Santis G, Osborne L *et al.* (1990) Linked marker haplotypes with the ΔF508 mutation in adults with mild pulmonary disease and cystic fibrosis. *Lancet*, **335**, 1426–1429.
Shoshani T, Augarten A *et al.* (1992) Association of a nonsense mutation (W1282X), the most common mutation in the Askenazi Jewish cystic fibrosis patients in Israel, with presentation of severe disease. *Am. J. Hum. Genet.*, **50**, 222–228.

CLONING OF THE MOUSE HOMOLOGUE AND MOUSE MODELS

Clarke LL, Grubb BR *et al.* (1992) Defective epithelial chloride transport in a gene-targeted mouse model for cystic fibrosis. *Science*, **257**, 1125–1128.
Gasparini P, Nunes V *et al.* (1991) High conservation of sequences involved in cystic

fibrosis mutations in five mammalian species. *Genomics*, **10**, 1070–1072.

Koller BH, Kim H-S *et al.* (1991) Toward an animal model of cystic fibrosis: targeted interruption of exon 10 of the cystic fibrosis transmembrane regulator gene in embryonic stem cells. *Proc. Natl Acad. Sci. USA*, **88**, 10730–10734.

Tata F, Stanier P *et al.* (1991) Cloning the mouse homolog of the human cystic fibrosis transmembrane conductance regulator gene. *Genomics*, **10**, 301–307.

Yorifuji T, Lemna WK *et al.* (1991) Molecular cloning and sequence analysis of the murine cDNA for the cystic fibrosis transmembrane conductance regulator. *Genomics*, **10**, 547–550.

THERAPEUTIC STRATEGIES

App E, King M *et al.* (1990) Acute and long term amiloride inhalation in cystic fibrosis lung disease: a rational approach to cystic fibrosis therapy. *Am. Rev. Respir. Dis.*, **141**, 605–612.

Knowles M, Church N *et al.* (1990) A pilot study of aerosolized amiloride for the treatment of lung disease in cystic fibrosis. *N. Engl. J. Med.*, **322**, 1189–1194.

Knowles M, Clarke L *et al.* (1991) Activation by extracellular nucleotides of chloride secretion in the airway epithelia of patients with cystic fibrosis. *N. Engl. J. Med.*, **325**, 533–538.

Mason S, Paradiso A *et al.* (1991) Regulation of transepithelial ion transport and intracellular calcium by extracellular adenosine triphosphate in human normal and cystic fibrosis airway epithelium. *Br. J. Pharmacol.*, **103**, 1649–1656.

Wagner J, Cozens A *et al.* (1991) Activation of chloride channels in normal and cystic fibrosis airway epithelial cells by multifunctional calcium/calmodulin-dependent protein kinase. *Nature*, **349**, 793–796.

GENE THERAPY

Rosenfeld M, Siegfried W *et al.* (1991) Adenovirus-mediated transfer of a recombinant α1-antitrypsin gene to the lung epithelium in vivo. *Science*, **252**, 431–434.

4 Huntington's Disease

DUNCAN J. SHAW

Finding the gene for Huntington's disease (HD) has long been one of the most sought after goals of human genetics research. This is because HD is such a devastating and untreatable disease, and because it shows straightforward autosomal dominant inheritance it was an ideal candidate for the application of positional cloning techniques. In this chapter, we look at the disease itself, how the gene was isolated, and how study of the mutation is beginning to explain some of the features of HD, as well as having many applications in the clinical situation.

HISTORICAL PERSPECTIVE

The term chorea, which comes from the Greek word κορσσ for 'dance', refers to the involuntary movements characteristic of HD and various other neurological disorders. Although choreic illness has been described by many observers from the Middle Ages onwards, it was the American family doctor George Huntington who in 1872 gave the first definitive description of the hereditary choreic disease that has now come to bear his name. His paper lucidly describes the essential features of the condition—its adult onset, progressive nature and fatal outcome, chorea and mental dysfunction, and autosomal dominant mode of inheritance. For a review of all aspects of HD the reader is referred to the book by Harper (1991). Undoubtedly some of the reports of similar conditions, from earlier in the nineteenth century, also describe HD. Going back even further one can find descriptions of 'St Vitus' dance' and 'dancing mania'; although similar in some respects to HD, these are likely to have had other causes. The persecution of sufferers from such conditions in mediaeval times was probably bound up with beliefs in witchcraft and diabolical possession, and has some recent parallels in the treatment of hereditary disease patients by the Nazis. Following Huntington's description, reports appeared describing HD in many other European and American countries, and Australia. Although HD is found in populations of non-European origin, it may well have been introduced there by European or American traders. Introduction of the disease gene by an outsider is

Molecular Genetics of Human Inherited Disease. Edited by D.J. Shaw
Published 1995 by John Wiley & Sons Ltd

certainly compatible with the existence of small isolated groups with exceptionally high disease prevalence, such as those around the shores of Lake Maracaibo in Venezuela. For these reasons, HD tends to be thought of as a disease of European origin.

CLINICAL ASPECTS

HD has a combination of neurological, psychiatric and genetic characteristics that together define the condition. Most patients present first with chorea or other neurological problems. Chorea is usually the most obvious aspect of more generalized motor dysfunction; other symptoms may include dystonia (sustained muscle contractions), rigidity (especially in the juvenile form of the disease), tics and myoclonus (brief, shock-like muscle jerks). Gait, eye movement, speech and ingestion of food may be adversely affected. Often in the later stages of the disease there is severe weight loss that is not responsive to diet. Psychiatric problems are virtually always present but tend to be non-specific; they include depression, irritability, schizophrenic symptoms, personality disturbance and dementia. One of the most important features of HD is its relentlessly progressive nature; this helps to distinguish it from conditions such as benign hereditary chorea. The average age at onset is in the early forties, but can range from childhood to old age. Juvenile-onset HD (age less than 20 years) is characterized by severe mental deterioration and rigidity rather than chorea. It is noteworthy that nearly all juvenile cases are of paternal origin, and that 'anticipation' (earlier onset in successive generations) is considerably greater in male than in female transmissions. Survival of patients from onset of symptoms to death is about 15 years, regardless of the age at onset. The patients usually die of pneumonia or cardiovascular disease, brought on by their progression to a state of immobility, weight loss, problems in taking nourishment, and general debility. Although some of the neurological and psychiatric symptoms can be partially treated with drugs, there is no satisfactory therapy, nor any way of curing the condition.

HD shows autosomal dominant inheritance with complete penetrance of the disease gene. In populations of European origin the prevalence of HD is about four to nine cases per 100 000, but because of the late onset of the disease the frequency of gene carriers is about three times higher than the prevalence. Most cases are familial; most so-called isolated or new mutation cases proved to have a family history when investigated further. Together with evidence from population studies, this suggests that there is a single (or small number of) mutational origin for all contemporary cases. Because onset is usually after child-bearing age, couples at risk cannot be sure they will not be affected and likely to pass on the disease gene to their offspring. The lack of useful pre-clinical markers for carriers of the HD gene has meant that one of the immediate implications of finding a DNA marker linked to the gene would be the possibility of offering pre-symptomatic diagnosis to individuals and couples at risk. Such procedures are now in routine clinical use, although the lack of a treatment or cure for the disease has discouraged the universal uptake of this service.

NEUROPATHOLOGY OF HD

Because the symptoms of HD are mostly motor and psychiatric, it is reasonable to assume that the underlying pathological process primarily affects the brain. Post-mortem studies on HD patients have confirmed this. The basal ganglia are the most obviously affected areas, with striking atrophy of the caudate nucleus and putamen, and a corresponding increase in the size of the lateral ventricles (Fig. 4.1). Loss of material occurs to a lesser extent in other parts of the brain. Post-mortem studies on brains of HD patients who died at various times during the course of the disease indicate that the degeneration is progressive. In the basal ganglia, it is predominantly the smaller-sized, spiny neurones that are lost (spiny refers to the appearance of the dendrites). Levels of γ-aminobutyric acid (GABA), one of the neurotransmitters found in spiny neurones, are reduced in HD brain, as are levels of met-enkephalin and substance P. However, another group of neurones, the aspiny interneurones, are relatively spared from degeneration. It also appears that the neurones most susceptible to degeneration in HD are those whose neurotransmitters rely on glucose metabolism for their synthesis, whereas those that are spared utilize transmitters whose synthesis is not dependent on glucose metabolism.

The observation that if the neurotransmitter glutamate remains in the synapse then death of the neurone will occur, has led to the development of the theory of excitatory neurotoxins as the agents of neuronal death in HD. Injection of the glutamate analogue kainic acid, or of the tryptophan metabolite quinolinic acid, into the striatal region of rat brain produces loss of neurones in a manner that has some similarities with HD. Some studies on the levels of metabolites of the quinolinic acid pathway in HD brain support this model, but other observations, such as the equivocal effects of the neurotoxins on neurones that are spared in HD, suggest that the theory should be treated with caution. In general, it appears that excitotoxicity is caused by the prolonging of the normal excitatory process, causing (for example) the opening of channels for calcium ions, and subsequent activation of proteases that cause intracellular damage.

THE SEARCH FOR THE HD GENE

Despite a great deal of work having been carried out on the pathology and biochemistry of HD, no coherent picture of the primary pathological process

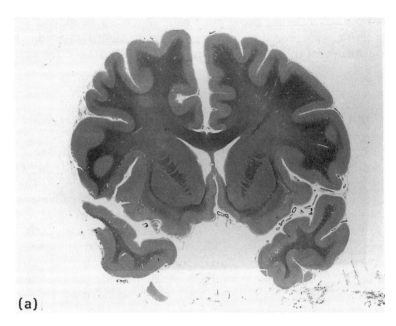

(a)

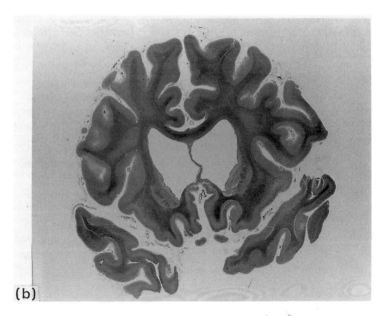

(b)

Fig. 4.1. Coronal sections through normal (a) and HD (b) brains. Enlargement of the ventricles due to death of neurones in the caudate nucleus can be seen. Courtesy of Dr. J. Neal, Department of Pathology, University of Wales College of Medicine

has emerged. In the early 1980s many investigators turned their attention toward the possibility of understanding HD by isolating the gene responsible. It was hoped that this might identify the type of protein that was defective in the disease, provide a basis for experiments to develop therapeutic strategies, and enable pre-symptomatic testing using DNA analysis to be carried out. It took about 10 years from the original discovery of linkage to the isolation of the gene itself (March 1993), and the mechanism of the HD mutation, expansion of a trinucleotide repeat sequence, was the fourth such example in a neurological disease (see chapters on myotonic dystrophy and fragile X syndrome in this book). The history of the hunt for HD illustrates the development and application of the many innovative technologies now in widespread use for positional cloning; however, it was the systematic application of a whole arsenal of methods, rather than any one particular technical breakthrough, that finally allowed the HD gene to be isolated.

Earlier attempts to find linkage to HD had to rely on the use of protein polymorphisms (blood groups, enzyme isoforms, etc.). These experiments are limited by the availability of such markers, which cover less than a third of the genome. Not surprisingly, no linkage was found. However, the growing realization in the early 1980s that the DNA itself could be used as a source of genetic polymorphism (Botstein et al., 1980), together with advances in molecular genetics that made the detection of such polymorphisms feasible, led to the early application of the recombinant DNA-based approach to the problem of HD. In order to search the genome systematically for linkage to a disease locus, it is necessary to have several hundred markers, fairly evenly spread so as to maximize the chance of at least one of them being close to the disease gene, together with DNA samples from clinically well-characterized families in which the disease is segregating. The first of these requirements was still far from being satisfied in the early 1980s, but progress was accelerating. Because HD is a relatively rare condition, and because its late onset makes it unusual for more than one generation of affected individuals in a family to be available for study, it can be difficult to assemble a panel of suitable family material for linkage analysis. The most remarkable effort in this direction was made by the Hereditary Disease Foundation, who ascertained and sampled a huge number of patients and their relatives, from the extended kindred living around Lake Maracaibo in Venezuela. Other investigators with a long-standing interest in HD were able to contribute by collection of DNA samples from families known to them.

Having assembled a panel of suitable family material, it was generally felt that the limiting factor in the search for linkage would be the time taken to work through a sufficient number of DNA markers—maybe several hundred. It was scarcely believable when Dr James Gusella at Harvard Medical School reported a significant linkage with one of the first dozen markers he tried. The probe named G8 detected a polymorphic locus (D4S10) on the short arm of chromosome 4 (Gusella et al., 1983). Further studies by other groups soon confirmed the linkage, and showed that the recombination fraction between HD and D4S10 was about 4%. The marker was localized using *in situ* hybridization and somatic cell analysis to the most terminal band on the short arm of chromosome 4, namely 4p16.3 (Fig. 4.2). Use of a second marker in this region, D4S62, showed that the HD gene was located distal to D4S10 and therefore within the 6 million bases (6 Mb) between it and the telomere. No evidence for the existence of more than one locus causing HD was found from any of the family studies.

Attention then focused on the isolation of new DNA markers from within 4p16.3, to provide a more accurate localization of the HD gene. These markers were isolated from libraries constructed either from flow-sorted chromosome

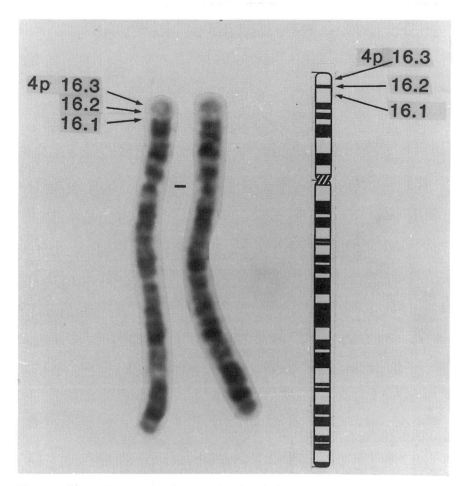

Fig. 4.2. Photomicrograph of part of a banded chromosome preparation, and diagrammatic representation (ideogram) showing locations of the first markers linked to HD on chromosome 4

4, or from somatic cell hybrids containing human chromosome 4 or fragments of it. The markers could be mapped relative to each other by genetic analysis in families, using reference pedigrees supplied by CEPH (Centre d'étude du polymorphisme humain) and by the recently developed technique of pulsed-field gel analysis. The latter technique allows one to construct restriction maps of chromosomal regions on the scale of several megabases, by a novel form of gel electrophoresis that separates very large DNA fragments. By a combination of these approaches, and the use of somatic cell hybrids containing various rearranged derivatives of chromosome 4, a genetic map of the markers on 4p16.3 was constructed (Fig. 4.3; Bucan *et al.*, 1990; Bates *et al.*, 1991). The polymorphisms detected by the markers were typed in the HD families, paying particular attention to cases where recombination had been detected between the HD gene and one or more markers. Unfortunately, this analysis did not point to a single unambiguous location for the HD gene. Rather, some of the recombination events indicated that the gene was located distal to all known markers, and therefore very close to the telomere, whereas others favoured a more proximal location in the region between D4S10 and D4S98 (Fig. 4.3; MacDonald *et al.*, 1989; Pritchard *et al.*, 1992). These two candidate regions (which will be referred to as the telomeric and proximal regions) were mutually exclusive, and the explanation for this conflict was not obvious.

At about this time there was a move afoot, involving most of the groups in Europe and the USA working on HD, to pool their resources and by eliminating competition in favour of cooperation to hasten progress toward isolation of the gene. This collaboration was sponsored by the Hereditary Disease Foundation in California. To try and resolve the problem of the two conflict-

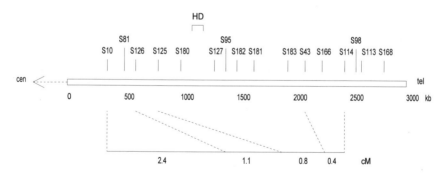

Fig. 4.3. Physical map of the terminal region of chromosome 4p, constructed by pulsed-field gel electrophoresis. Above the open box are shown positions of DNA markers used for linkage analysis (S10 = D4S10, etc.). Distances in kilobases are shown below the box. Below this are shown the genetic distances (in cM, derived from family studies) of selected markers. The non-linearity of genetic and physical distance in this region is illustrated. HD indicates position of the gene eventually identified as the Huntington's gene

ing gene locations, members of the collaborative group decided to take two different approaches. The first of these was to isolate the DNA between the most distal marker and the telomere, and see whether this contained genes that might be candidates for HD. This was achieved by construction of a hybrid yeast–human artificial chromosome, with one of its telomeres derived from human chromosome 4 and the other being of yeast origin (Fig. 4.4). Analysis of the DNA in this clone showed that it was comprised mostly of repetitive sequences, and did not contain anything resembling conventional genes (Youngman *et al.*, 1992). The second approach was to refine the genetic localization of HD, using linkage disequilibrium analysis. Because HD is a disease with a very low rate of new mutation, it was reasoned that for markers extremely close to the disease gene most contemporary HD families might

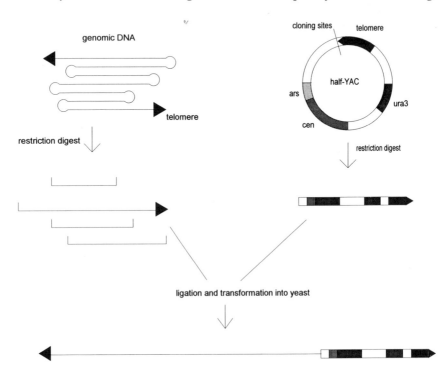

Fig. 4.4. Isolation of a human telomere by complementation in yeast. Genomic DNA was digested to large fragments with an infrequently cutting restriction enzyme and ligated to the 'half-YAC' vector, which supplies all the requirements for a yeast artificial chromosome except for one telomere. After transformation into yeast, selection for the ura3 marker enables clones to be recovered in which a DNA fragment containing a human telomere is joined to the vector. The human and vector sequences are not drawn to scale (human sequence is much longer than vector). Features of vector are: ars, autonomously replicating sequence; cen, centromere; ura3, uracil biosynthesis gene for selection in a ura⁻ host strain. From Bates *et al.* (1990), with permission

retain alleles characteristic of the chromosome on which the mutation first arose. This would be revealed by analysis of large numbers of normal and HD chromosomes. If the allele frequencies for a particular marker are significantly different in the population of disease chromosomes, as compared to the normal population, this is evidence for that marker being in linkage disequilibrium with the disease. This is expected to occur only for those markers that are closest to the disease locus, as recombinations will have occurred since the original mutation event between less close markers and the disease, resulting in loss of linkage disequilibrium. When this analysis was carried out, it was found that linkage disequilibrium did exist, but only for those markers in the proximal candidate region (D4S95 and D4S98; see Fig. 4.4; Snell et al., 1989; Theilman et al., 1989). Because similar results were presented by three independent laboratories, and since there was no evidence for an HD candidate gene in the telomeric region, the search for the HD gene was now concentrated almost entirely on the proximal candidate region.

Further linkage disequilibrium analysis indicated that it was not possible to predict exactly where in the 2 Mb candidate region the gene would be, although a 500 kb segment around the markers D4S180 and D4S182, which showed the most strongly conserved haplotype of marker alleles in HD chromosomes, was slightly favoured by some workers (Fig. 4.3). So it was deemed necessary to assemble a set of cloned DNAs encompassing the whole region. This was achieved by a collaborative effort, using cosmid and YAC (yeast artificial chromosome) libraries constructed from flow-sorted chromosome 4, from somatic cell hybrid DNA, and from total genomic DNA. Clones were characterized by restriction mapping and by use of specific probes from the by now well-mapped candidate region, and sets of overlapping clones or 'contigs' were assembled (Baxendale et al., 1993).

The contigs of genomic clones were next used to find candidate genes. Because it was not apparent what kind of gene HD would be, any transcribed sequence that mapped within the correct region would be have to be assessed as a candidate. However, because of the predominantly neurological nature of the disease, most investigators felt confident in using brain cDNA libraries as the source of expressed sequences. Although it is possible to find candidate cDNAs by probing cDNA libraries with fragments of individual clones from the candidate region, this approach is very time-consuming when a region of 2 Mb has to be searched. Therefore, considerable effort was put into finding ways to speed this process up. Two such methods were of particular value for HD.

The first of these involved using a whole YAC to probe cDNA libraries. Since the YACs for the HD region were up to 700 kb in length, this approach offered the possibility of covering the whole region with a small number of hybridizations. However, the complexity of YAC DNA used as a probe means that large amounts of radioactivity have to be used in the hybridization and the conditions must be very well judged for the experiment to be successful. Furthermore, contamination of cDNA libraries with yeast gene sequences is a common problem, and these clones give positive signals in the hybridization experiment and have to be eliminated at a later stage. Despite these problems, many useful cDNAs were isolated by YAC screening (Snell *et al.*, 1993a).

The second method for rapid detection of expressed sequences has become known as 'exon trapping' or 'exon amplification' (Buckler *et al.*, 1991). In this procedure, fragments of genomic DNA are subcloned into a vector that contains a promoter, part of a known gene, and a transcription terminator (Fig. 4.5). If this vector is introduced into a mammalian cell line in tissue culture, it will produce an RNA transcript of known length. However, if the DNA fragment cloned into the vector contains one or more exons of a gene, with associated RNA splice sites, then the RNA produced will be of increased size since it will also contain the spliced exon(s) of the foreign sequence. By analysing the RNA from the cell lines by reverse transcription and polymerase chain reaction (PCR), clones containing putative exons, and therefore representing genes, can be recovered. The method can be accelerated by analysing pools of clones in the initial stages, and then subcloning to individual clones any pools that show a positive signal. The method has the advantage of not relying on cDNA libraries as the primary source of expressed sequences, and so can in principle find genes that are not represented in the cDNA libraries that one has access to.

These approaches led to the isolation of a number of candidate genes, including some that correspond to known proteins (a fibroblast growth factor receptor, cGMP phosphodiesterase β-subunit, and α-adducin, a protein believed to play a role in organization of the cytoskeleton). However, none of these genes showed alterations specific to HD and could therefore be ruled out as candidates, as could a number of other genes with no known homology that had been isolated from the candidate region.

THE HD GENE AND MUTATION

One of the precepts of the HD collaborative research group was to avoid duplication of effort by working on the same cDNAs or particular segments of the candidate region. The team at Massachusetts General Hospital, led by Drs James Gusella and Marcy MacDonald, had chosen to focus on the 500 kb segment around the markers D4S180 and D4S182, where linkage disequilibrium was most pronounced (Fig. 4.3). Using exon trapping of cosmids from this interval, they had identified parts of a large gene called IT15 ('interesting transcript number 15'). The 5' end of this gene proved difficult to isolate, but eventually five cDNA clones were recovered which together defined a gene with a transcript of about 11 kb (Huntington's Disease Collaborative Research Group, 1993). The sequence of these clones included an open reading frame of

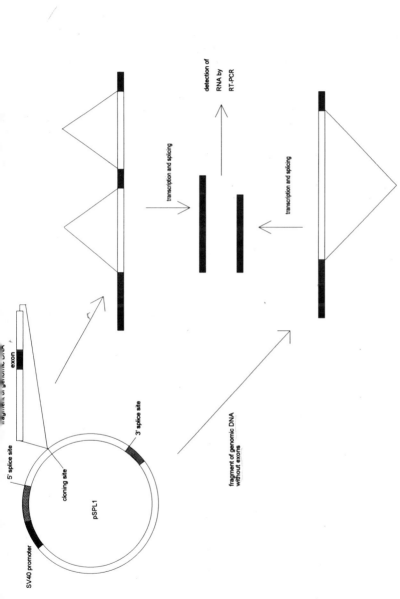

Fig. 4.5. Isolation of human exons in cloned DNA by 'exon trapping' (or 'exon amplification'). The pSPL1 vector contains the following features: SV40 transcriptional promoter (dark shaded box); exons derived from the *b*-globin and HIV *tat* genes (light shaded boxes); intron of the HIV *tat* gene (open box, bounded by 5' and 3' splice sites). After cloning fragments of human DNA into a site within the intron, vectors are transfected into mammalian cells in culture (cos cells) and after a short period of growth RNA is harvested. Analysis by reverse transcriptase PCR, using primers derived from the vector, allows the presence of exons within the cloned DNA fragments to be detected as these give rise to longer RNA products

9432 bases, encoding a polypeptide of 3144 amino acids with a predicted molecular weight of 348 kDa. However, the most remarkable feature of the sequence was a run of 21 tandemly repeated copies of the trinucleotide repeat CAG, located near the 5′ end of the gene and within the predicted open reading frame. The potential significance of such a trinucleotide repeat sequence was obvious, in view of recent discoveries (see other chapters in this book) involving the genes for fragile X syndrome, myotonic dystrophy and Kennedy's disease. A second copy of the IT15 sequence was obtained, this time from a cosmid derived from an HD chromosome, and shown to have not 21 but 48 copies of the CAG repeat. A PCR assay for the repeat was designed and applied to DNA samples representing 74 HD and 173 normal chromosomes. The results showed that all the normal chromosomes had numbers of repeats in the range 11–34, whereas all the HD chromosome repeat numbers were in the range 42 to about 100 (Figs 4.6 and 4.8). This established beyond doubt that the expanded trinucleotide repeat was the causative HD mutation, and indicated that disruption of the normal function of the gene IT15 was the most likely cause of the primary biochemical abnormality. Unfortunately, the predicted sequence of the IT15 protein (named 'Huntingtin') did not resemble that of any known protein, and so its exact function remains a mystery, at least for the time being. Northern blot analysis showed that the gene was expressed in most tissues, including the brain. In situ hybridization to RNA showed that the transcript was found in all parts of human and rat brains, with higher expression in neuronal than glial cells. Similar results were obtained with an antibody to Huntingtin, which showed staining in the cytoplasm of all cell types, and also in the nuclei of neuronal cells. The gene is subject to alternative poly-adenylation, such that a 13.7 kb transcript predominates in brain, and a 10.3 kb message is more abundant in other tissues. The difference between these mRNAs seems to reside in the lengths of their 3′ untranslated regions.

PROPERTIES OF THE HD MUTATION

Although the function of the HD protein had not immediately been revealed by cloning the gene, rapid analysis of the mutation was possible by several laboratories, which helped to illuminate some of the puzzling aspects of HD genetics. Those individuals in which the HD gene seemed to have recombined with all the markers, pointing to a telomeric gene location, were shown not to be carrying the mutation at all. The simplest explanation is that these patients were misdiagnosed as HD. The existence of HD homozygotes, which had been suspected from linked marker studies in rare families with both parents affected, was confirmed by mutation analysis. These patients were no more severely affected than heterozygotes, and their cells still expressed IT15

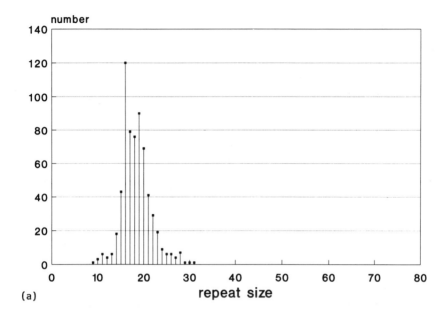

(a)

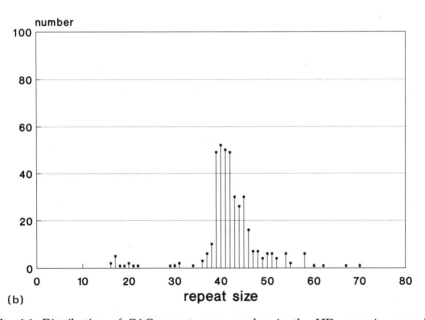

(b)

Fig. 4.6. Distribution of CAG repeat copy number in the HD gene, in normal chromosomes (a) and HD chromosomes (b). The HD chromosomes with repeats less than about 35 proved on further investigation to be spurious. Reprinted with permission from Snell *et al.* (1993b)

mRNA, showing that the effect of the mutation is not to cause a loss of gene function.

Shortly after the description of the HD gene was published, three papers appeared with data on the characteristics of the mutation, based on a combined data set of over 1200 HD and 2200 normal chromosomes (Duyao *et al.*, 1993; Snell *et al.*, 1993b; Andrew *et al.*, 1993). The range of repeat copy numbers is between 9 and 37 in normals, and (with a few exceptions) between 37 and 121 in HD chromosomes (Figs 4.6 and 4.8). The exceptions are apparently affected individuals with repeat sizes in the normal range. Although many of these are likely to be due to misdiagnosis or other types of error, the possibility remains that some of these may have other explanations, including different mutations in the HD gene, mutations in other genes, or modifying effects on the expression of the HD gene. There are also a small number of individuals in an intermediate size range (34–38 repeats) who may be at risk of passing on an increased expansion to their offspring. The size of the repeat is significantly correlated with the age at onset, so that patients with the largest expansions tend to have the earliest onset (Fig. 4.7). Most parent-to-child transmissions of the HD gene result in a small change (in either direction) in the size of the repeat, with the largest increases found in male transmissions. The sperm of affected males was found to contain larger and more heterogeneously distributed repeat sizes than their somatic tissue, which may explain the tendency for the offspring of affected males to

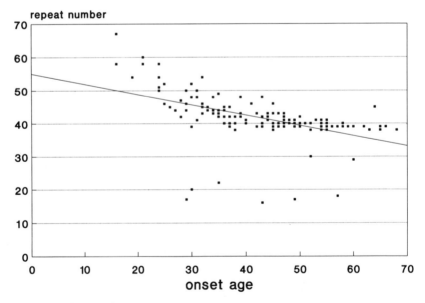

Fig. 4.7. Correlation of the repeat number in HD chromosomes with age at onset. The isolated points well below the line correspond to the spurious points shown in Fig. 4.6(b). Reprinted with permission from Snell *et al.* (1993b)

83

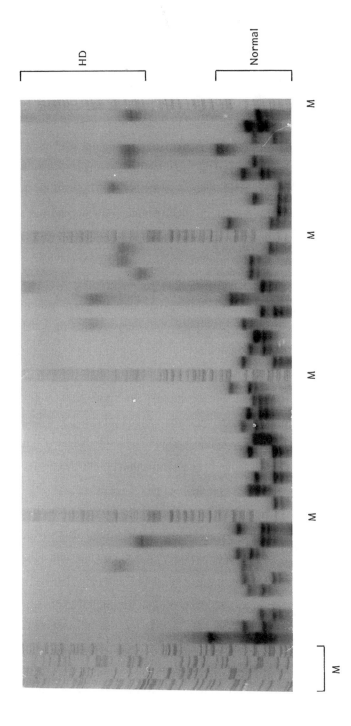

Fig. 4.8. The Huntington's disease CAG repeat, detected by PCR followed by acrylamide gel electrophoresis and autoradiography. Lanes marked M contain size standards. The range of repeat copy numbers for normal and HD alleles is shown on the vertical axis

have increased repeat sizes relative to their fathers. The phenomenon of anticipation, whereby age at onset is earlier in the offspring than in the parent, was confirmed at the molecular level, including the observation that this effect is most pronounced in male transmissions. Finally, the repeat size of the *normal* allele was weakly but significantly correlated with age at onset, but only when the father was the unaffected parent. It has been reported that the father's age has an effect on the age at onset of HD in the offspring, even when the father is the unaffected parent. Age-related, progressive changes in the activity of the normal HD allele, as a result of somatic instability or alterations in methylation, may be the underlying cause of these phenomena. Their restriction to the paternally derived chromosome suggests that some form of genetic imprinting is involved.

Around 20 cases of HD with no family history were found to carry expanded repeats typical of HD; they had unaffected parents with repeat sizes of 30–38, at the upper end of the normal range. It was proposed that instability of the sequence had resulted in a slight size increase during transmission from parent to child, across the threshold of the disease range. In all cases where the parental origin could be determined, the father was the transmitting parent, suggesting that instability occurs during male gametogenesis. This result is in agreement with the finding that the sperm of affected males contains generally larger and more heterogeneously sized expansions, although corresponding data for oocytes have not been obtained.

The CAG repeat shows moderate instability in transmission from HD parents to offspring. Most transmissions result in a change in expansion length, with those transmitted maternally showing increases or decreases of up to three repeats, and paternal transmissions giving rise to changes of −2 to +28 repeats. This is in marked contrast to the behaviour of the trinucleotide repeats in myotonic dystrophy and fragile X syndrome, which often show dramatic expansions of up to several thousand copies. But the repeats in both of the latter diseases are within untranslated regions of the corresponding genes, whereas in HD the repeat is probably in the coding sequence and translated as glutamine. It may be that selection acts at the cellular level against larger HD expansions, due to complete disruption of protein function, and that this prevents the repeats in HD from sustaining a size of more than about 100 copies. In HD it appears from studies on sperm that expansion of the trinucleotide repeat takes place during male gametogenesis (results for female gametogenesis are not available). The repeat also appears to be stable during subsequent mitotic cell divisions, since there is no variation between tissues of an individual, and identical twins have identical repeat lengths. This finding is in contrast to that in fragile X, where expansion of the repeat sequence seems to occur in a specific window in early fetal development (see Chapter 5, this volume).

The diseases in which the trinucleotide repeat is within the coding sequence of the gene (HD, Kennedy's disease, spinocerebellar ataxia 2, and dentato-rubral-pallidoluysian atrophy) contrast markedly with those where the repeat is untranslated (myotonic dystrophy, fragile X) in terms of the degree of repeat expansion that is observed. In the former case, normal alleles contain up to about 30–40 repeats, and disease alleles up to about twice this figure. In the latter, repeats in affected patients can increase to a copy number of several thousand. The former group of diseases, which are all neurological in nature, may share a common underlying disease mechanism where an excessive number of glutamine residues in the affected protein causes dysfunction (for example, progressive degeneration) in those neuronal tissues characteristically affected.

CLINICAL APPLICATIONS

Ever since the linkage with D4S10 was discovered, it has been possible to perform a pre-symptomatic test for the HD gene, at least in families with affected members available for study. This linkage-based approach had many drawbacks. The need for the whole family to be involved, and to cooperate in giving blood samples, was often hard to satisfy in practice. The implications of a positive result in the consultand for other family members who did not request testing needed to be considered, as did the possibility of a change in predicted clinical status for relatives, following the onset of symptoms in a previously asymptomatic individual. These and other clinical and ethical issues are described in detail by Harper (1991).

Now that the mutation responsible for HD in the majority of cases has been found, many of these problems will disappear. Testing may now be done with a DNA sample from the consultand alone, with no implications for other family members. The PCR method that is used is reasonably robust in practice, and the only problem in interpretation is likely to come from the few individuals who have repeat numbers of about 37, at the threshold between the disease and normal ranges. These individuals may not themselves be affected but may be at increased risk of passing on a larger, disease-causing allele to their offspring (see above). A more accurate PCR assay has now been developed, which avoids inclusion in the PCR product of a polymorphic CCG repeat found close to the disease-associated CAG repeat. This increases the accuracy of the test by up to three repeat units. However, although the laboratory aspects of the test have been improved, the need for proper counselling and follow-up for HD patients and their families is still as great as ever, and protocols will have to revised in the light of this latest advance. Testing of any individual at risk for HD, who has not specifically requested to know their status, must be avoided at all costs. DNA samples from such individuals, and from families whose pedigree structure did not allow prediction using linked markers to be done, is often stored in the laboratories now able to carry out mutational analysis. Testing of these must not be done

except in response to an informed request from a properly counselled patient.
The specificity and sensitivity of the mutation test suggests that it will be of
use in differential diagnosis. The diagnosis of HD can be a problem, especially
in isolated cases, and confusion with other choreic syndromes (hereditary or
otherwise) and inherited neurodegenerative diseases sometimes occurs.
Specific testing for the HD mutation will be a valuable addition to the
armoury of diagnostic procedures.

FUTURE PROSPECTS

At present, nothing is known of the normal function of the IT15 gene product,
nor of how the HD mutation interferes and causes the symptoms of the
disease. Studies are now being done to characterize the distribution of gene
expression (using *in situ* hybridization to tissue sections) and of the protein
product (using antibodies raised to peptides derived from the primary
structure of the protein). These experiments may provide some clues to
possible functions for the gene product, and can also be carried out on
samples from HD patients to answer the question whether variation in the
repeat number of the CAG sequence affects tissue or intracellular distribution. Both alleles of the gene appear to be expressed in heterozygotes, and
quantitative aspects of the effect of the mutation on levels of RNA or protein
can be further addressed using reverse transcriptase PCR or immunoprecipi-
tation and western blotting, respectively. At present it is not certain whether
the repeat sequence is within the coding sequence of the gene and translated
into protein, although it is generally assumed that it is. It should be possible to
address this by purification of the protein using immunoaffinity chromato-
graphy and determining the N-terminal amino acid sequence. A possible
molecular basis has been proposed for HD and the other diseases (such as
Kennedy's disease) where an expanded CAG repeat would cause expansion
of a poly-glutamine sequence in the protein product. The enzyme trans-
glutaminase is responsible for cross-linking of proteins such as involucrin in
keratinocytes, forming an insoluble matrix. If by expansion of the poly-
glutamine sequence the HD protein were to become a substrate for trans-
glutaminase, this would result in formation of aggregates with possible
adverse consequences for the cell. Such an effect is expected to have a
dominant phenotype, and if it were a relatively slow process, the late
age-at-onset of the disease could be accounted for. Although it is simplest to
assume that the symptoms of HD are caused by an effect of the CAG repeat on
the protein (Huntingtin) predicted from the IT15 gene sequence, there are
other possibilities, such as effects on other transcripts located next to or
within introns of the gene.
    In myotonic dystrophy and fragile X syndrome, there is strong evidence
from linkage disequilibrium analysis that most affected individuals are

descended from one or a very few initial mutation events. As we have seen in this chapter, the same is true of HD. In the case of the few apparent new HD mutations, the chromosomal background on which these have arisen has the same haplotype as that most commonly associated with familial cases, suggesting that it predisposes the CAG repeat to undergo expansion into the disease range. If the most common cause of 'new' mutations in HD is expansion of an allele from the upper end of the normal range, into the disease range, then one would expect to find the largest-sized normal alleles associated with the chromosomal haplotype characteristic of HD chromosomes. This could be investigated by studying linkage disequilibrium between the CAG repeat and nearby markers in normal chromosomes. Similar studies that have been done for myotonic dystrophy support the view that the source of mutations in that disease is larger-sized normal alleles that undergo gradual expansion into the disease-causing range; such a mechanism may apply to trinucleotide repeat disease in general. The very low prevalence of HD in some populations (such as native Africans) may be due to a lack of individuals with repeats in the high normal range. This prediction could be tested by appropriate population studies.

There are no naturally occurring animal models for HD, but now that the gene has been isolated one could be created. The cloned mouse homologue of the gene could be mutated *in vitro* in various ways, so as to introduce extra CAG repeats to mimic the situation in an HD patient, or to knock out gene function altogether (for example, by introducing a premature stop codon). The mutations would then be introduced into mouse embryonal stem cells by homologous recombination, transferred to blastocysts by microinjection, and transplanted into a female mouse. The chimeric offspring can be bred from to produce pure transgenic lines, which can then be crossed with each other to construct homozygotes and various types of heterozygotes. The effects of these mutations on the neurological functioning of the animals could then be studied. Transgenic animals will also be a critical resource for the testing of possible therapeutic agents.

The predicted sequence of the gene product shows no similarities with known structures, and protein biochemistry and structural biology are not yet at the stage where a three-dimensional structure or function for a novel sequence can be predicted. The function of the poly-glutamine stretch is not clear: however, this type of feature is found also in a number of transcription factors such as *Drosophila* homeobox proteins, where it appears to be essential for normal activity. In order to investigate whether the HD protein could be involved in transcriptional regulation, DNA binding studies using purified protein could be undertaken.

Apart from the poly-glutamine sequence, other noteworthy features include two poly-proline stretches just downstream, and a leucine zipper motif that could be involved in dimerization. Therefore, possible roles of the protein can only be guessed at, taking into account what is known of the disease pathology. As well as transcriptional regulation, areas to investigate would include energy metabolism (because of the characteristic weight loss), neurotransmitter toxicity, and premature neuronal ageing and death, possibly via disruption of the normal mechanisms controlling apoptosis. The ultimate aim of HD research is a complete understanding of the disease process at the cellular and molecular levels, with a view to therapeutic intervention. These goals are likely to be brought about only by application of the skills of a wide variety of biomedical scientists, but it is fair to say that the molecular genetic analysis, culminating in isolation of the HD gene and mutation, was the critical first step in this endevour.

REFERENCES

Andrew SE, Goldberg YP, Kremer B et al. (1993) The relationship between trinucleotide (CAG) repeat length and clinical features of Huntington's disease. Nature Genet., 4, 398–403.
Bates GP, MacDonald ME, Baxendale S et al. (1990) A yeast artificial chromosome telomere clone spanning a possible location of the Huntington disease gene. Am. J. Hum. Genet., 46, 762–775.
Bates GP, MacDonald ME, Baxendale S et al. (1991) Defined physical limits of the Huntington disease gene candidate region. Am. J. Hum. Genet., 49, 7–16.
Baxendale S, MacDonald ME, Mott R et al. (1993) A cosmid contig and high resolution restriction map of the 2 megabase region containing the Huntington disease gene. Nature Genet., 4, 181–186.
Botstein D, White RL, Skolnick M and Davis RW (1980) Construction of a genetic linkage map in man using restriction fragment length polymorphisms. Am. J. Hum. Genet., 32, 314–331.
Bucan M, Zimmer M, Whaley WL et al. (1990) Physical maps of 4p16.3, the area expected to contain the Huntington's disease mutation. Genomics, 6, 1–15.
Buckler AJ, Chang DD, Graw SL et al. (1991) Exon amplification: a strategy to isolate mammalian genes based on RNA splicing. Proc. Natl Acad. Sci. USA, 88, 4005–4009.
Duyao M, Ambrose C, Myers R et al. (1993) Trinucleotide repeat length instability and age of onset in Huntington's disease. Nature Genet., 4, 387–392.
Gusella JF, Wexler NS, Conneally PM et al. (1983) A polymorphic DNA marker genetically linked to Huntington's disease. Nature, 306, 234–238.
Harper PS (ed.) (1991) Huntington's Disease. London: WB Saunders.
Huntington's Disease Collaborative Research Group (1993) A novel gene containing a trinucleotide repeat that is expanded and unstable on Huntington's disease chromosomes. Cell, 72, 971–983.
MacDonald ME, Haines JL, Zimmer M et al. (1989) Recombination events suggest possible locations for the Huntington's disease gene. Neuron, 3, 183–190.
Pritchard C, Zhu N, Zuo J et al. (1992) Recombination of 4p16 DNA markers in an unusual family with Huntington disease. Am. J. Hum. Genet., 50, 1218–1230.
Snell RG, Lazarou LP, Youngman S et al. (1989) Linkage disequilibrium in Huntington's disease: an improved localisation for the gene. J. Med. Genet., 26, 673–675.
Snell RG, Doucette-Stamm LA, Gillespie KM et al. (1993a) The isolation of cDNAs within the Huntington disease region by hybridisation of yeast artificial chromo-

somes to a cDNA library. *Hum. Mol. Genet.*, **2**, 305–309.

Snell RG, MacMillan JC, Cheadle JP *et al.* (1993b) Relationship between trinucleotide repeat expansion and phenotypic variation in Huntington's disease. *Nature Genet.*, **4**, 393–397.

Theilman J, Kanani S, Shiang R *et al.* (1989) Non-random association between alleles detected at D4S95 and D4S98 and the Huntington's disease gene. *J. Med. Genet.*, **26**, 676–681.

Youngman S, Bates GP, Williams S *et al.* (1992) The telomeric 60kb of chromosome arm 4p is homologous to telomeric regions on 13p, 15p, 21p, and 22p. *Genomics*, **14**, 350–356.

5 The Fragile X Mental Retardation Syndrome

DAVID L. NELSON

ABSTRACT

The mutation involved in the fragile X mental retardation syndrome was the first of a new class of mutations involving simple repeat sequences disturbing the function of genes in which they reside. The repeat in the fragile X-related mental retardation gene (FMR1) is composed of the triplet CGG, which is found to be unstable in unaffected carriers and can undergo dramatic expansions in length, leading to perturbation of gene transcription and phenotypic effects. The behaviour of this repetitive element has clarified many of the odd genetic features of fragile X syndrome, and provided a new perspective on long-recongized peculiarities in human genetics such as penetrance, expressivity and anticipation. This perspective has been strengthened considerably with the discovery of similar repeats with many of the same properties in five other human genetic disorders. The function of the FMR1 gene is as yet unknown; however, it is likely to be the sole gene involved in the phenotype. Efforts by several groups led to the identification of the fragile X region and gene and involved some of the standard methods of positional cloning as well as some techniques specific to the fragile X locus.

THE FRAGILE X SYNDROME PHENOTYPE

Fragile X syndrome (McKusick #30955, Martin–Bell syndrome) is the most frequently encountered form of inherited mental retardation in humans, with a frequency estimated to be between 1 in 1000 and 1 in 2500 and is second only to trisomy 21 as a genetic cause of mental deficiency. The phenotype is found in both males and females, reflecting the X-linked dominant nature of the disorder. However, penetrance is incomplete, with 80% of males and 30% of females showing impairment (Sherman *et al.*, 1984). The phenotype is also highly variable, reflecting variable expression of the mutation.

Molecular Genetics of Human Inherited Disease. Edited by D.J. Shaw
Published 1995 by John Wiley & Sons Ltd

MENTAL RETARDATION

The hallmark of the fragile X phenotype is mental retardation. The degree of mental impairment in fragile X patients can range from mild to profound. Table 5.1 shows the wide variability in mental ability among affected fragile X males and females. In affected males, mean IQ scores in a variety of studies have been found to cluster in two regions. Boys, teenagers and young adult males with fragile X syndrome have a mean IQ of ~50, while studies of older individuals give means of 35. These results are often interpreted to reflect a decline in IQ with age; however, the possibility that these reflect differences due to intervention must be considered, since the younger patients are less likely to have been institutionalized, and may therefore represent a higher functioning class. Long-term longitudinal studies of a single fragile X population should allow resolution of this issue; two such studies have concluded that the effect is authentic. It remains to be seen whether this post-pubertal decline in IQ represents a real decline, or decreased learning potential. In females, there is less profound retardation, with borderline to mild encompassing the majority of the affected females. Newer data suggest that approximately one-half of females carrying the fully mutated allele are mentally retarded; however, based on cytogenetic and now careful molecular analysis, it is becoming clear that some intellectual deficit (IQ depression of ~20–30 points) is found in most female carriers of full mutations. These studies have been complicated by the difficulties in identifying female carriers by the cytogenetic assay; with the use of the molecular assays, it should become more straightforward to identify larger cohorts of known carriers of various types to determine the mutation's effect on IQ.

BEHAVIOURAL ASPECTS

A comprehensive discussion of the type of mental retardation and the range of behavioural effects of fragile X syndrome is beyond the scope of this

Table 5.1. Variability in mental retardation. Fragile X mental retardation is highly variable. IQ ranges and the percentage of fragile X males and females who fall into each range are shown. Data taken from the table on page 115 of Sutherland et al. (1985).

Males		Females	
Borderline	8%	Borderline	56%
Mild	12%	Mild	24%
Moderate	50%	Moderate	19%
Severe	20%	Severe	<1%
Profound	8%		

chapter. However, some of the features often ascribed to fragile X males include poor eye contact, hand flapping, perseverative speech with echolalia and palilalia (self-repetition), hyperactivity with attention deficit, aggressive outburst and many features typical of autism.

PHYSICAL FEATURES

Males with fragile X syndrome can exhibit a constellation of physical features attributable to the disorder. However, none of these features is completely diagnostic, as no one physical feature is found in 100% of affected individuals. Also, many of the physical features overlap with other mental retardation syndromes, so that a differential diagnosis based solely on physical examination is difficult. Many of the features appear to derive from a connective tissue defect; these include large and prominent ears, velvet-like skin, hyperextensible finger joints, high arched palate, flat feet, pectus excavatum and mitral valve prolapse. In addition, a distinctive long and narrow face with prominent jaw is present in many males. Perhaps the most useful finding in post-pubertal males is macroorchidism (large testicles). These are found in ~80% of fragile X males after puberty; however, in younger boys the percentage is much lower (~15%). Female carriers with mental retardation occasionally show some of the facial features found in males, but in general their physical characteristics are within the normal range.

PATHOLOGY

Limited post-mortem studies of neuroanatomy have revealed no significant findings in brains of fragile X individuals. One magnetic resonance imaging study of living patients found decreased volume in the posterior vermis of the cerebellum. Enlarged testicles are found to have oedema and some increased connective tissue (collagen), but are otherwise normal. Fragile X males produce viable sperm and have fathered children, although this is rare due to the mental retardation. Ovarian enlargement has been observed in some female carriers, indicating that the gonadal effect may occur in both sexes.

GENETICS OF FRAGILE X SYNDROME

Inheritance of fragile X syndrome is quite complicated, although the features of the mutation at the DNA level (see below) provide an explanation for the unusual genetics found in this disorder. The most striking aspect of fragile X is its incomplete penetrance in both males and females with the mutation. This is particularly interesting in the case of 'normal transmitting males' (NTMs) who transmit the mutation to grandsons but are unaffected themselves. More complicated are the probabilities of mental deficiency based on

94   D.L. NELSON

the affected status of relatives. This has become known as the 'Sherman paradox' (Sherman *et al.*, 1984), where the probability of mental retardation is increased by the number of generations through which the mutation has passed, and is higher for both sons and daughters of affected females as well as for females with affected sibs. Fig. 5.1 illustrates the Sherman paradox with an imaginary pedigree. Note the increasing likelihood of mental retardation as the generations progress. Once affected individuals are found, the probabilities become close to those expected from Mendelian genetics; however, in the prior generations, significantly reduced probabilities are found.

The Sherman paradox encompasses some features of anticipation, although it does not fit the strictest definition of the term (worsening symptoms or earlier age of onset in subsequent generations). Anticipation is usually described in terms of variable expressivity rather than incomplete penetrance as in fragile X syndrome. However, since the probability of being affected increases in subsequent generations, anticipation has been applied to fragile X inheritance as well.

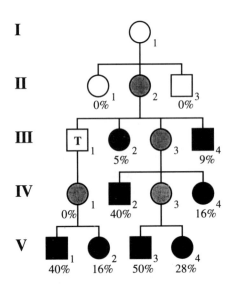

Fig. 5.1. The Sherman paradox. This figure represents a composite pedigree to illustrate risks of mental retardation based on position within the pedigree. Risks are given in per cent below each figure. Fully shaded figures indicate mental retardation, while partially shaded figures are unaffected carriers. The male designated with a T is a 'normal transmitting male', who carries a fragile X chromosome without phenotypic effect

CYTOGENETICS IN FRAGILE X SYNDROME

FRAGILE SITES: ARTEFACTS OF CELL CULTURE CONDITIONS

The history of fragile X syndrome is an interesting one. Originally described by Martin and Bell in a large English pedigree (Martin and Bell, 1943), this type of X-linked mental retardation with macroorchidism was referred to as Martin–Bell syndrome for nearly 30 years. In 1969, Lubs described a 'marker X' in families with X-linked mental retardation (Lubs, 1969); this aberrant X chromosome showed a peculiar constriction at the distal end of the long arm (in the Xq27.3 band). The marker X was lost to analysis shortly thereafter, as the rapidly changing cell culture media of the time were altered to types that no longer supported visualization of the fragile site. It was several years later when Sutherland, exploring various media, defined reproducible conditions under which the marker chromosome could be seen. A small percentage of metaphase nuclei showed loss of Xq28 by breakage, leading to the term 'fragile site'. Numerous common and rare fragile sites on the other human chromosomes and in other species were subsequently identified; however, the site at Xq27.3 (FRAXA—for the first fragile site on the X chromosome) is the only one with a firmly established associated phenotype. Despite much speculation, there is no evidence that the fragile site is expressed as such *in vivo*, nor that breakage and loss of the Xq28 band occur.

FRAXA is one of the 20 folate-inducible fragile sites. Additional fragile sites are characterized by other chemical induction systems. Conditions which induce FRAXA in cells in culture appear to do so by altering the mechanics of DNA replication or repair. The primary nucleotide pools affected by fragile site induction conditions are dCTP and dTTP.

DIAGNOSTIC APPLICATION OF THE FRAGILE SITE

Once reproducible conditions for expressing the fragile site were established, cytogenetics became the primary diagnostic tool for defining fragile X syndrome. While results were rather variable among laboratories, a major source of confusion resulted from inconsistencies in the visualization of the fragile site, ranging from no expression to high and low percentage of positive metaphases in members of the same family. Prenatal diagnosis was challenging due to low levels of expression in amniocytes and chorionic villi and the requirement of fetal blood sampling for reliable diagnosis. With the current knowledge of the molecular genetics of fragile X syndrome, the variability of expression of fragility is somewhat easier to understand, since the mutation is hypervariable within individuals and between family members. Also, two new, very closely linked fragile sites (FRAXE and FRAXF) have been identified which have been found in families with mentally retarded members previously misclassified as fragile X mutations. With the capability of performing DNA-based testing for fragile X syndrome (see below), the role of cytogenetics will be altered to ruling out other cytogenetic abnormalities.

THE HUNT FOR THE FRAGILE X GENE

Three groups independently arrived at the region of the X chromosome involved in cytogenetic fragility in the spring of 1991. These were the groups of Jean-Louis Mandel of INSERM, Strasbourg; Grant Sutherland and Rob Richards of Adelaide Children's Hospital in Adelaide (with assistance from David Schlessinger at Washington University); and an American/Dutch consortium, composed principally of the laboratories of C. Thomas Caskey and David Nelson at Baylor College of Medicine, Houston, Stephen Warren at Emory University, Atlanta and Ben Oostra of Erasmus University, Rotterdam.

Efforts by several groups to identify the fragile X site by the method of positional cloning date back to the early 1980s. Isolation and linkage analysis of DNA fragments in or near the Xq27.3 band in which the fragile site is observed led to a number of new markers in the vicinity. Among these was the St14 probe defining a variable-number tandem repeat sequence at the DXS52 locus, first described by I. Oberlé and still one of the best genetic markers in the Xq28 band. It was quickly apparent that clinical fragile X syndrome co-segregated genetically with the polymorphic markers near the fragile X site. Some still unresolved linkage results were obtained with markers near the haemophilia B (clotting factor IX) locus, with some families showing significant crossover frequencies and others showing none. Nonetheless, a set of markers was found to be useful for following fragile X chromosome in pedigrees and for prenatal diagnosis.

Physical mapping in the region employed a number of X chromosome breakpoints in Xq27 and q28. These were used to position DNA markers with respect to one another and to support the generation of additional markers to establish the genetic location of fragile X syndrome in the region. Three developments led to the eventual isolation of the fragile site and the FMR1 gene. These were the isolation of fragile X chromosomes broken at (or very close to) the fragile site, establishment of large-insert cloned DNAs derived from the region, and identification of large restriction fragments whose sizes were specifically altered in fragile X males due to methylation. Each of these developments is described in detail below. Fig. 5.2 illustrates many of the features described in the following sections with a map of the Xq27.3 region surrounding the fragile X site.

SPECIFIC TRANSLOCATIONS AT THE FRAGILE SITE

Somatic cell hybrids retaining a fragile X chromosome were produced by Warren in the mid-1980s. These were originally important in establishing that

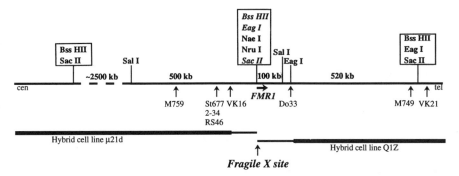

Fig. 5.2. The fragile X region. Restriction enzyme sites are indicated above the line; those in boxes are present in CpG islands and are subject to methylation. The middle box contains the three sites (italics) found in pulsed-field gel studies to be specifically methylated in fragile X males. Probes are indicated with arrows below the line. Two representative hybrid cell lines are shown for the panel of hybrids with breakpoints at the fragile site. Two representative hybrid cell lines are shown for the panel of hybrids with breakpoints at the fragile site. Distances are as indicated

induction of the fragile X site was not dependent upon other genetic factors in autosomal locations, as hybrids retaining the fragile X as their only human material still exhibited the fragile site after induction. Warren used the observation that fragility occasionally resulted in chromosome breakage to select for chromosomes broken and translocated to hamster chromosomes. These studies took advantage of the flanking hypoxanthine phosphoribosyl transferase (HPRT) and glucose 6-phosphate dehydrogenase (G6PD) genes for selection and screening respectively. This led to a panel of hybrid cell lines retaining either the distal or proximal portions of the X and lacking the reciprocal part (Warren *et al.*, 1990). Cytogenetically, the breakpoints were very close to the fragile site, and many of the hybrids retained their ability to induce fragility at the site of translocation. These panels of hybrids later were shown to have breakpoints within the fragile X repeat in the majority of cases, and were critical reagents in identification of the fragile X mutation in each of the laboratories involved.

METHYLATION ANOMALIES NEAR THE FRAGILE SITE

As probes were identified close to the fragile site (defined by hybrid cell mapping panels as well as by genetic distances), pulsed-field gel electrophoresis was used by several groups to generate maps of the region. One particular hope was that altered fragment sizes would be identified in the translocated chromosomes derived in the experiments of Warren (Warren *et al.*, 1990). Such a result was eventually found by the American/Dutch consortium using probes derived from closely linked yeast artificial chromosomes (YACs) (Verkerk *et al.*, 1991). Prior to that, however, both the Mandel and Davies groups reported identification of methylation abnormalities in fragile X patients detected on pulsed-field gels using probes distal to the fragile site (Vincent *et al.*, 1991; Bell *et al.*, 1991). Mandel's group described analysis of 21 patients and 22 controls, using the enzymes *Bss*HII, *Eag*I and *Sac*II and the probe Do33, a randomly chosen Alu polymerase chain reaction (PCR) product derived from a radiation-reduced hybrid cell line.

Fragile X males showed aberrantly large bands, much larger than the 620 kb fragment found in normal males and females, but similar to the large bands seen in female cell lines. This suggested the possibility of a methylation anomaly in the region, leading to the inability of these methylation-sensitive enzymes to cleave. This hypothesis was consistent with the concept of local X-inactivation having an effect in the syndrome (Laird, 1987).

The Davies group found similar results using a probe derived from YACs identified by microdissection experiments aimed at developing probes closely linked to the fragile site (Bell *et al.*, 1991). Anomalies with *Bss*HII and *Sac*II were observed by this group as well, and were also interpreted as being caused by increased methylation rendering some sites uncleavable. This was very well demonstrated by the change in pattern from grandfather to grandson, with no mutation visible at this scale, but simply increased methylation. The CpG island exhibiting unusual methylation was thus identified by both groups, and this became the target of chromosome walking to the presumed location of the fragile X defect.

YEAST ARTIFICIAL CHROMOSOMES

A major development assisting in the pursuit of the fragile site was the establishment of the YAC cloning system and of libraries containing clones from the region surrounding the fragile X mutation. First described by Burke, Carle and Olson, YACs have developed into essential tools for the identification and characterization of large genomic regions. They retain DNA fragments ranging in size from 100 to over 1000 kb in length; this is a 2.5 to 25-fold improvement over the largest convenient cloning vectors (cosmids) available previously. Largely due to the intrinsic interest in fragile X syndrome and other disease loci in the distal portion of the human X chromosome, the region around the fragile site was targeted by several groups for cloning in YACs. Both the Houston and St Louis groups used a somatic cell hybrid retaining Xq24-qter as its only human component (this chromosome was originally derived from a fragile X patient) in a hamster background to construct YACs specifically derived from this region. Other groups utilized libraries of total human DNA to isolate YACs from Xq27 and q28 with markers known to reside in these regions. Finally, unsuccessful attempts were made by at least three groups to identify cosmids and YACs containing the junctions of translocations involving the fragile site by identification of clones carrying both human and hamster DNA.

Considerable difficulty was encountered with isolation of YAC contigs (contiguous regions isolated in overlapping YAC clones) covering the interval between the closest flanking markers. This was due to a combination of bad luck (two YACs identified from the Centre d'Etude Polymorphisme Humain (CEPH) library with a close proximal marker languished in the Mandel group for nearly a year, and were only later found to have distal ends a few tens of kilobases away from the closest known distal flanking marker, thus demonstrating that they flanked the fragile site) and to inherent instability (or 'unclonability') of this region of human DNA carried in yeast cells (both the St Louis total human and X-specific libraries were devoid of clones spanning the fragile site, as was the Houston X-specific library). YACs found in the ICRF library tended to extend to the fragile site, but not across it. Cosmids from the region were apparently under-represented in a number of libraries as well.

One YAC clone was found in the St Louis collection that allowed the Sutherland group to identify the fragile site. This clone was contained in a library derived from the X3000-11 cell line containing a fragile X chromosome isolated in a hamster background. The clone, XTY26, was generated in an attempt to isolate the Xq telomere. This library was made with the 'half-YAC' approach, where the mammalian telomere is asked to provide telomere function in the YAC (see Fig. 4.4). Only one of the two vector arms is used to create such clones. XTY26 was isolated using the proximal probe VK16 (Yu et al., 1991) and found to be a circular YAC. It is particularly significant as it is the only clone to contain an expanded, full mutation allele, although it has deleted a substantial fraction of the repeats and flanking sequences (Kremer et al., 1991).

YACs crossing the fragile site were also identified in the library produced by the Centre d'Etude du Polymorphisme Humain (CEPH) by the Mandel and the American/Dutch consortium independently. Both groups identified YACs 209G4 and 141H5 which derive from the normal male used to produce the YAC library. The American/Dutch consortium used a small YAC (RS46) derived from one of the X3000-11 libraries to identify the larger overlapping clones (Verkerk et al., 1991), while the Mandel group identified the clones with the random probe St677 (Heitz et al., 1991) and later learned that the fragile site had been crossed when the distal probe Do33 (used to identify methylation anomalies) was developed.

Demonstration that YACs spanned the fragile site took two routes. The primary one was use of the somatic cell hybrids with breakpoints presumed to be in the fragile site. Generation of probes contained in the same YAC which spanned the fragile X specific breakpoints in hybrids retaining either proximal or distal portions of the X chromosome was accomplished by all three groups (Heitz et al., 1991; Verkerk et al., 1991; Yu et al., 1991). The Mandel and Sutherland groups also reported fluorescence in situ hybridization data (FISH) demonstrating that YACs (and smaller clones) spanned the cytogenetic fragile site as a further demonstration of the appropriate location (Heitz *et al.*, 1991; Yu *et al.*, 1991). Similar data using cosmids were present in the original communication from the American/Dutch consortium (Verkerk *et al.*, 1991); however, these were removed prior to publication, and were later published in a review from the Oostra group.

IDENTIFICATION OF THE FRAGILE SITE

With YACs spanning the fragile site as defined cytogenetically and by the somatic cell hybrid breakpoints, and with the clue of aberrant methylation of a CpG island, all groups zeroed in on cloning the island from the YACs. Three papers described the isolation of clones containing the island and their use in analyses of fragile X individuals (Oberlé *et al.*, 1991; Verkerk *et al.*, 1991; Yu *et al.*, 1991). Cosmid or lambda subclones of the YACs were used by each group to walk to the CpG island. Sites in the CpG island showed the expected specific methylation in fragile X males which had been found initially by pulsed-field gel studies, and appeared to be close to or at the site of breakage in the Warren hybrids. However, the more intriguing aspect of the DNA fragments spanning the fragile site was the finding of length variation within fragile X families, with hypermutability of the fragile X alleles. Furthermore, the sizes of fragments containing the region of the CpG island correlated both with methylation and affected status in males (Oberlé *et al.*, 1991; Yu *et al.*, 1991). Bands of increased size showed a higher propensity to methylation, and methylation was the hallmark feature of affected males. Fig. 5.3 shows a restriction map of the region immediately surrounding the fragile X site.

Two general classes of mutation were seen. These have become known as the 'pre-mutation' and 'full mutation' and were first defined by increases in the sizes of fragments observed in Southern hybridization. Pre-mutations are found associated with all NTMs and many carrier females, and involve increases in the length of this region by 50–500 bp. Full mutations are found in all affected individuals, male or female, and some carrier females. The full mutation alleles show increases of 600–3000 bp in length and are usually heterogeneous within the individual sample, demonstrating significant somatic variation of the mutant allele. The specific methylation increases are observed in full mutations, but are not found in pre-mutation alleles. Fig. 5.4 illustrates these two types of mutation.

CGG REPEATS CONFER INSTABILITY

The Mandel and American/Dutch consortiums both reported identification of a sequence of CGG trinucleotide repeats in the region of instability and methylation (Oberlé *et al.*, 1991; Verkerk *et al.*, 1991), and proposed their involvement in the hypermutable behaviour of the region. This suspicion was directly demonstrated by the Sutherland group (Kremer *et al.*, 1991), which

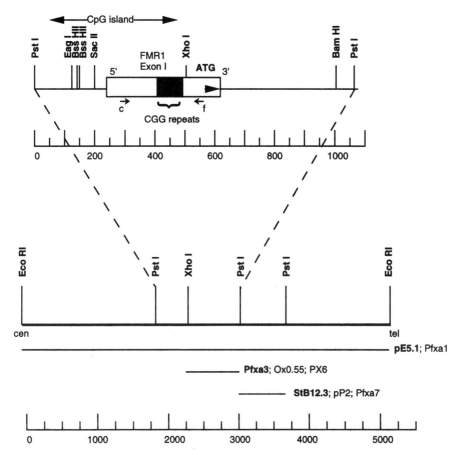

Fig. 5.3. The fragile X site. A restriction site map of the 5200 bp *Eco*RI fragment surrounding the fragile X site derived from a normal X chromosome is shown below, with locations and names of some of the probes used in fragile X diagnostics indicated as lines below. In the upper part of the figure, the 1060 bp central *Pst*I fragment containing the fragile site, the CpG island and the first exon of FMR1 is shown to scale. Restriction sites are as indicated; the FMR1 exon is shown as a box, with shading to represent the location of the normal 30 CGG repeats. The ATG start of translation codon is indicated, as is the 5' to 3' orientation of the transcribed mRNA. Arrows marked 'c' and 'f' are the primers used for PCR assays across the fragile site to determine the repeat number

showed that the repeats were the site of DNA alterations in fragile X chromosomes, that they were polymorphic in the normal population, that two of the Warren hybrids had broken within the repeats, and that the flanking sequences were unchanged between normal and fragile X chromosomes. Additional studies by the American/Dutch consortium (Fu *et al.*, 1991) characterized the behaviour of the repeats in greater detail (see below).

IDENTIFICATION OF THE FMR1 GENE

The American/Dutch consortium reported the isolation of a cDNA from the region of the fragile X mutation, derived from a gene designated FMR1 (for fragile X mental retardation 1). The cDNA was found using entire cosmids from the region as probes in a human fetal brain cDNA library. One cosmid containing the lengthy 3' untranslated region of the gene identified an initial cDNA (BC22) which was used to identify an additional cDNA (BC72). BC72 and BC22 combined to form a 3.7 kb sequence with an open reading frame extending 1975 bp from one end of BC72. Included in the sequence were the CGG repeats found at the methylation-sensitive CpG island in genomic DNA. This indicated that an exon of the FMR1 gene contained the hypermutable sequence at the fragile site. No other genes have yet been identified in this region. Numerous considerations now point to the primary role of FMR1 in generation of the fragile X phenotype (see below); however, the normal function of FMR1 is as yet undetermined. A review of the current knowledge of FMR1 and its characteristics is provided below.

THE FMR1 GENE

Prior to the identification of the fragile site and CGG repeat mutation, there was significant speculation regarding the nature of the fragile X mutation's effect on linked genes and the disease phenotype. Serious hypotheses included the possible extinction of expression of several genes due to altered chromatin, mosaic loss of Xq28 sequences or inappropriate X-inactivation. The observation of mutations within an exon of a gene with an expression pattern consistent with phenotype features of the disease offered significant evidence of its role in the disorder. The evidence has mounted in the interim, and coupled with the absence of other genes in the region has led to the current conclusion that FMR1 is the sole gene involved in the fragile X phenotype.

FMR1 CHARACTERISTICS IN NORMAL INDIVIDUALS

The FMR1 gene produces a transcript of approximately 4.4 kb, which can be found in most tissues by northern hybridization. It is particularly abundant in brain and testes RNA samples, concordant with the pathology found in fragile X patients (Hinds *et al.*, 1993). Studies of expression by *in situ* hybridization to mouse tissues show high levels of mRNA in early embryogenesis, as well as high levels in specific structures in brain (hippocampus and granular layer of the cerebellum), oesophagus and testes (Hinds *et al.*, 1993) (Table 5.2). A similar study of human embryos concluded that the hippocampus and nucleus basalis magnocellularis were the major sites of FMR1 expression in developing brain (25 weeks' gestation) (Abitbol *et al.*,

Table 5.2. FMR1 RNA expression in
adult tissues. Relative levels of mRNA
expression in a variety of adult murine
tissues is given. Data are adapted
from Hinds et al. (1993).

Brain	+++
Testis	+++
Oesophagus	++++
Thymus	++++
Spleen	+++
Ovary	+++
Eye	+++
Colon	++
Uterus	++
Thyroid	+
Liver	+
Kidney	+
Lung	+
Heart	−
Aorta	−
Muscle	−

1993). Recent data suggest a transiently high level of FMR1 in murine spermatogonia, concomitant with development of spermatids and suggesting a role of FMR1 in sperm development.

Sequence analysis of FMR1 has not allowed prediction of function (Verkerk *et al.*, 1991). Other than the CGG repeat sequence found at the 5' end of the mRNA, no sequences in FMR1 demonstrate significant similarities with known genes in database analyses. In addition, nothing in the predicted peptide sequence suggests function due to motifs or domains. Thus the function of FMR1 is yet to be determined, and will require examination of the protein's location and interactions. Studies of the protein are beginning to be reported, and the initial findings suggest a primarily cytoplasmic localization, with tissue abundancies similar to those of the mRNA (Verheij *et al.*, 1993). These studies will likely extend the observations of mRNA abundance and begin to suggest functional roles by cellular localization.

Multiple isoforms of FMR1 protein are predicted by the observation of extensive alternative splicing. Some of the alternative mRNA forms predict altered reading frames and termination codons, suggesting quite different protein types. Four different isoforms were detected in the initial report of protein analysis (Verheij *et al.*, 1993), and these likely correspond to some, but not all of the predicted isoforms. Again, the function(s) of the different isoforms awaits analysis at the polypeptide level.

The FMR1 gene consists of 17 exons spanning approximately 40 kbp of Xq27.3 and is transcribed in a proximal to distal orientation. The first exon contains the CGG repeats and the translational start codon 69 bp downstream of the repeats (Fig. 5.3). There is a large first intron, but other features are unremarkable.

The FMR1 mRNA is highly conserved among vertebrates, and homologous sequences have been observed by hybridization using Southern analysis in species as divergent as yeast, *Caenorhabditis elegans* and *Drosophila* (Verkerk *et al.*, 1991). The human and murine sequences are 97% identical at the peptide level (98% similarity). The CGG repeat sequences are also found in the murine gene in a similar position; however, they are fewer in number (~10) and relatively invariant among strains (Nelson, unpublished). All mammalian species studied demonstrate the presence of CGG repeats (Nelson, unpublished), suggesting a functional role for these sequences.

FMR1 IN FRAGILE X SYNDROME

The most interesting feature of the original cDNA clones described for FMR1 was the trinucleotide repeat sequence found at the 5' end of the open reading frame. This sequence was of the form $(CGG)_9AGG(CGG)_9AGG(CGG)_{10}$, and was later shown to be the site of mutation in fragile X families (Fu *et al.*, 1991; Kremer *et al.*, 1991). Mutations take the form of small (pre-mutation) or quite large (full mutation) increases in the length of this repeat, immediately identifying FMR1 as having a primary role in fragile X pathogenesis. Since the reading frame was open through these repeats to the 5' end of the cDNA, it was thought possible that these were in protein coding sequence, and could encode a contiguous run of arginine residues whose increase in number would result in a mutant protein. Several lines of evidence now exclude the repeats from the coding sequence. These include the presence of upstream stop codons in extended cDNA sequence, and reduced similarities between human and mouse sequences 5' to an ATG start codon compared to 3' to that codon (Fig. 5.3). In addition, identically sized polypeptides can be demonstrated in normal and pre-mutation individuals with widely variant numbers of repeats (Verheij *et al.*, 1993).

The mechanism by which repeat expansion leads to fragile X syndrome involves methylation-mediated loss of transcription of the FMR1 mRNA. The aberrantly sized fragments in *Bss*HII digests observed on pulsed-field gels by Mandel's group resulted from methylation of a CpG island located 200 bp 5' to the repeats, in the promoter region of the FMR1 gene. Expansion of the repeats correlates with methylation at the island (Fig. 5.4).

Complete loss of expression of FMR1 RNA in 80% (16/20) of fragile X males studied was found by the American/Dutch consortium (Pieretti *et al.*, 1991). Four cases with RNA expression demonstrated partial methylation of the CpG island, and three of these were mosaic, containing both methylated full and unmethylated pre-mutation fragments. The fourth case with under-methylation and RNA expression was a long-term lymphoblastoid cell line.

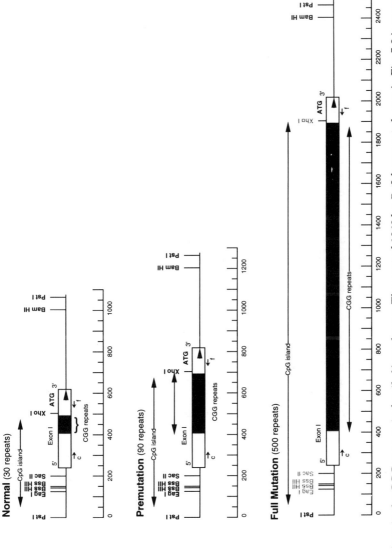

Fig. 5.4. Mutations in the CGG repeats in fragile X syndrome. The central 1060 bp *Pst*I fragment shown in Fig. 5.3 is repeated in the top portion of the figure (see legend of Fig. 5.3 for details). The middle portion shows this region from a pre-mutation of 90 repeats, to demonstrate the alterations in the size of the repeats and the exon. The bottom portion shows the region in a full mutation of 500 repeats. In this panel, restriction sites subject to fragile X specific methylation (which results in failure to cleave) are shown in grey

The 16 cases with no expression showed complete methylation in *Bss*HII digestion and no evidence of pre-mutation-sized fragments. The four patients with undermethylation were not detectably less severely affected (and this has been the general case with patients exhibiting mosaic mutations). However, given the somatic mosaicism in the mutation, it is possible that the blood cells studied are irrelevant to the phenotype and that the methylation status is different in affected tissues. Fragile X syndrome is unique among human genetic diseases in that it is the only disorder to date known to be due to aberrant methylation.

It is yet unclear at what point in development the methylation and loss of FMR1 expression occur. Some insight into this question can be obtained from the study of chorionic villus (CV) and amniocentesis samples obtained for prenatal diagnosis. The majority of male CV samples have been found to be undermethylated when carrying the full mutation, and two examples have been shown to have no methylation despite a large full mutation. In one of these latter cases, FMR1 mRNA was detected, indicating both that the full mutation has no effect on transcription in the absence of methylation and that FMR1 expression is not necessarily missing from conception onwards. Another potential lesson from this study regards the sequence of events in extinction of FMR1 expression. At least in the extra-embryonic cells of the chorion, the potential for normal levels of FMR1 exists.

The observation of unmethylated full mutations suggests that the mutation is inherited in an unmethylated state and becomes methylated during development. For the alternative hypothesis to be valid, i.e. that the mutation is inherited in an 'imprinted' (which could be interpreted as a methylated) state and remains shut off (lingering X-inactivation (Laird, 1987)), would require postulating that methylation is lost in the extra-embryonic lineages. It is as yet uncertain which of these possibilities is correct. However, it is tempting to speculate that some of the variability seen in the fragile X phenotype could be due to variability in the timing of onset and location of methylation. In the study of FMR1 transcription in the unmethylated CV sample, assay of fetal tissue obtained at a gestational age of 13 weeks demonstrated nearly complete methylation with some lingering FMR1 mRNA. Thus it may be the case that FMR1 is variably expressed during development of an affected fetus, and that this could play a role in the wide range of phenotypic features.

Approximately 50% of full mutation-bearing females are found to be mentally retarded. This is presumed to be due to skewing of X-inactivation. The FMR1 gene is subject to X-inactivation in normal females. In affected females, no significant correlation between blood-based skewing of inactivation and mentation can be demonstrated. However, the relevant structures in the brain may show such skewing if they were assayed. The wide variability in degree of mental ability in females is also likely due to biased inactivation in a tissue-dependent manner, in addition to the suggested variation in onset of methylation in fetal development.

Another intriguing possibility is that the fragile X mutation-induced methylation spreads, leading to loss of expression of a number of adjacent genes as well, and leading to a contiguous gene 'shutoff' mutation. Since no other genes have been found within 200 kb on either side of the fragile site, and since methylation has not been found to spread beyond this, this has become an unlikely hypothesis. In addition, the identification of individuals with a fragile X-like phenotype without the CGG repeat mutation has established FMR1 as the main gene involved in the syndrome.

FMR1 MUTATIONS NOT INVOLVING THE CGG REPEAT

Two FMR1 deletions have been reported. Each of the patients involved is reported to have a fragile X-like phenotype, which is well within the range of variation of patients carrying the CGG repeat mutation. One patient exhibited a deletion of ~2.5 million base pairs, including the entire FMR1 gene and 2 Mbp proximal as well as several hundred thousand base pairs distal. This patient is reported to be moderately retarded and to fit the classic fragile X diagnosis. The second patient found to exhibit a deletion shows macroorchidism, mild retardation and fragile X-like facies. His deletion ablates a region from the middle of FMR1 extending 250 kb proximal, including the CGG repeats and the CpG island. These data suggest that no other essential genes are likely to reside in the 2.5 Mb region, that complete loss of expression of FMR1 is no more deleterious than the potential 'partial' loss of expression from methylation repression, and that any other gene whose mutation is necessary for the phenotype must reside in the 250 kb region proximal to FMR1 that is deleted in common in the two patients.

A third FMR1 mutation has been found involved in a fragile X-like presentation; however, this is a point mutation, and it results in a rather more severe phenotype. This patient is severely mentally retarded, with an IQ below 20, and exhibits profound macroorchidism and focal seizures. This latter feature is uncommon in fragile X syndrome. A single base change was identified in position 367 of the FMR1 mRNA. This mutation would predict an amino acid change of Ile to Asn at this position. This is a new mutation in the patient; it is not found in his mother's DNA. This patient's disease suggests strongly that FMR1 is the gene involved in fragile X syndrome. The increased severity of the phenotype (and the extension of the symptoms) suggests that this mutant FMR1 protein is more deleterious than absence of the protein. It is also possible that such a mutation would act in a dominant fashion; however, since the patient's mother does not exhibit the mutation, it is not possible to conclude that this is truly a dominant negative mutation. Such a conclusion awaits further study of the protein and this mutant version.

FMR1 AND TREATMENT

As the common mutation in fragile X syndrome involves methylation-induced repression of transcription, it is conceivable that treatment with demethylating agents could reinduce expression of the gene and have an ameliorative effect. The observation of mRNA production from full mutation alleles without methylation encourages such attempts. Much more detailed study of the role of FMR1 in development will be required to understand the deficits in fragile X patients. Since the clinical features are present from birth, it is likely that lack of FMR1 expression during pre-natal development causes irreversible damage, and this notion is supported by the early and abundant levels of FMR1 mRNA observed during fetal development (Abitbol et al., 1993; Hinds et al., 1993). However, this is as yet uncertain and will require a much deeper understanding of the normal function of FMR1. Speculation regarding the potential for gene therapeutic approaches to fragile X syndrome will likewise await further understanding of FMR1's function and the possible consequences of its improperly regulated expression.

DNA INSTABILITY AND FRAGILE X GENETICS

The identification of unstable DNA at the fragile X site provided part of the answer to the puzzle of genetic inheritance in fragile X syndrome. The studies of Sherman et al. (1984) had pointed out the increasing likelihood of retardation in fragile X pedigrees in subsequent generations. This 'Sherman paradox' could not be explained by normal Mendelian transmission genetics. Another puzzling observation was that only offspring of female carriers were at risk, and that the disorder required transmission through a female for there to be an effect. The behaviour of the CGG repeats at the fragile site allowed resolution of some of these issues.

SOUTHERN ANALYSIS

Using Southern blots, each of the initial reports of the fragile site showed unusual behaviour of the mutant DNA fragments. The 'full' mutations associated with affected individuals were often mosaic, i.e. hybridization was observed over a broad size range in the gel. This is the result of mitotic instability (although the question of when the instability occurs is an interesting one—see below). Intergenerational instability was also seen, with the most impressive examples being the large increases associated with the transitions from pre- to full mutations found when some carrier mothers transmitted to their affected offspring. These can result in 10 to 20-fold increases in the repeat number (2–5 kb increases in fragment size). The most interesting feature of these large expansions, relevant to the genetics of the disease, was their maternal-specific nature. In no case were full mutations found in daughters of males carrying pre-mutations. This provided a molecular explanation for the observation of a maternal transmission requirement of the disorder; however, there is still no clear mechanistic explanation. It was also apparent by Southern analysis that the pre-mutations were unstable intergenerationally. Even when remaining pre-mutation sized (increases from normal by less than 600 bp), these often were found to differ in size between parent and child. The variation was confirmed to be due to the CGG repeats with use of smaller probes and PCR of flanking sequences (Kremer *et al.*, 1991); however, the estimate of the mean number of repeats and their variability was inaccurate due to the use of a deleted clone for sequence analysis and to the inability to amplify DNA containing the repeats. Nonetheless, most of the behaviour of the fragile X mutations was worked out using Southern analysis.

The estimated number of repeats required for a phenotypic effect is ~230; however, this threshold is not absolute. Alleles of this size can be methylated or not, and this appears to be the factor critical to phenotypic effects. In general, affected individuals are found to have restriction fragments suggestive of several hundred repeats—well above this threshold. Alleles below 200 repeats usually confer no obvious phenotype.

PCR ANALYSIS

The precision of PCR-based methods offered the opportunity for better estimation of the repeat numbers and their variation. However, the region offered significant difficulties due to the high GC content of the DNA. A method to amplify through the repeats was developed and reported in late 1991 (Fu *et al.*, 1991) (Fig. 5.4). The method provided analysis of the repeat lengths in normal and pre-mutation individuals; however, it was unable to amplify through full mutation alleles. An enhanced PCR method was later reported that additionally allowed detection of PCR products in the full mutation alleles. While this latter method provides significant improvement for diagnostic applications, it has not yet led to increased understanding of the fragile X mutational mechanism.

CGG REPEAT INSTABILITY

PCR has allowed the number of repeats at the fragile site to be determined with precision in normals and fragile X pre-mutation carriers. A range of sizes has been found in the normal population, supporting the observation of polymorphism seen in Southern analysis. The most common allele has been measured by acrylamide gel electrophoresis at 29 repeats; however, it may actually contain 30 repeats. Sizes ranging from 6 to 54 repeats were observed in a random collection of normals (Fu *et al.*, 1991), with fragile X pre-mutations found as small as 52 repeats and ranging upwards to 200, with the majority in the 80–120 interval. One hallmark of the repeats in fragile X families was their completely unstable character. In all instances where pre-mutations were transmitted from parent to child, the number of repeats was found to change. This extended the observations from Southern analysis and showed for the first time the extraordinary high mutation frequency of the fragile X pre-mutation alleles. Individuals with somatic variability were also identified, indicating mitotic instability as well. In 75 transmissions of normal-length alleles, no size changes were observed.

The overlap between normal and pre-mutation allele sizes has grown since this original report, with fragile X families exhibiting alleles as small as 43 repeats and normal individuals as high as 60 (Nelson, unpublished). Among the high-end normal alleles, some have been found to exhibit instability in germ-line transmission (Fu *et al.*, 1991). It is possible that these represent fragile X mutations in their earliest stages, and the differences between these alleles and those of similar length showing stable transmission will likely prove important to understanding the triggers of allele instability.

RISK OF EXPANSION TO THE FULL MUTATION

Through PCR analysis of mothers' allele sizes in a large number of fragile X families, it was possible to determine that risks of expansion from pre- to full mutations increased with the length of the pre-mutation (Fu *et al.*, 1991). No alleles below 60 repeats expanded to full mutations when transmitted, while all alleles above 90 repeats were found to expand. Intermediate allele lengths showed increasing risks of expansion, with those in the 60s having less than 20% risk, while those in the 70–90 range were closer to 80% (Table 5.3).

Table 5.3. Risk of pre- to full mutation in female transmission. Risks dependent upon repeat number are shown for five ranges of repeat sizes. Risks were determined by observation of the number of full mutation offspring divided by the total number of offspring receiving the mutant chromosome. Data are adapted from Fu et al. (1991).

Repeat number	Risk of expansion to full mutation
<60	<1%
61–70	17%
71–80	71%
81–90	86%
>91	>99%

Moreover, the pre-mutation lengths tended to increase in each subsequent generation of any given family. These two observations can account for the increasing risks of retardation seen in subsequent generations. The length of the repeat increases and increased length increase the risk of expansion to the fully mutated form, resulting in disease. These results have been validated by other groups in additional families, and adequately explain the Sherman paradox. However, it is still unclear why the expansions are limited to female germ-like transmission.

WHY FEMALE-SPECIFIC EXPANSION?

The mechanism by which the CGG repeats undergo dramatic expansion is completely unknown. In addition, it is unclear at what point in development this 10 or 20-fold increase in repeat number might happen. All observations rely on accessible tissues in mother and child, and these have typically been blood or cell lines of lymphoblastoid origin. The alterations in the repeats could happen at any point in the production of oocytes, or even post fertilization. Some recent studies begin to shed a faint light on these events.

Full mutations have been found to show considerable mosaicism. This was initially interpreted to be due to ongoing mitotic instability of the expanded CGG repeats. However, two cases of full mutation-bearing twins indicated that this instability is likely occurring during embryonic or fetal development after which the allele is largely stabilized. Recent work studying long-term culture of cells carrying a full mutation confirms its stable nature; the mutation can be stably maintained in clonal cell lines after as many as 25 doublings (Wöhrle et al., 1993). These data together suggest that the mosaicism is established early in development and is then stabilized. While other scenarios can be envisaged, one possibility is that the mosaicism is established by multiple, post-conceptional transitions of a pre- to full mutation. If correct, this model must account for the maternally contributed X-specificity of these transitions, as the paternally contributed X does not exhibit expansions.

Another intriguing result was found through analysis of sperm from four full mutation-affected fragile X males (Reyniers et al., 1993). These males' sperm exhibited only pre-mutation alleles, despite full mutations present in blood of each. This can be interpreted as evidence for post-conceptional expansion, occurring exclusively in somatically destined cells, with the germ-line having been spared. An alternative explanation could be selection in spermatogonia for FMR1 protein production and reduction of the full mutation. This latter view is supported by recent evidence of FMR1 production in developing spermatogonia. A selection in sperm could also explain the inability of the male germ-line to transmit full mutations. This might account for the observation of female-only generation of full mutations, as those which occur in the male germ-line are incapable of production of viable sperm. Resolution of these competing theories awaits the direct analysis of oocytes from pre-mutation women.

ORIGINS OF THE FRAGILE X MUTATIONS

Due to its very high population frequency, and to the fact that the phenotype is usually a genetic lethal in males, it was anticipated that an extraordinarily high rate of new mutations would be found in fragile X syndrome. Since estimates of the frequency of affected males in the general population ranges from 1 in 1000 to 1 in 2000, the new mutation frequency required to replenish the lost alleles would be on the order of 1 in 3000 to 1 in 6000. These are the highest new mutation frequency estimates for any human genetic disorder. However, no new CGG-based mutations have been found in any of the fragile X families assayed. In every family studied, there is a pre-existing pre-mutation as far back as the pedigree can be traced. Additionally, some present-day fragile X individuals have been found to be distantly related, with common ancestors dating back to the mid-1700s. These ancestors can be presumed to have carried at-risk alleles. This suggests very strongly that the fragile X pre-mutation is not deleterious, and that it can be carried for dozens of generations prior to exhibiting disease. Thus 'new' mutations may reflect recent increases into the size range at risk for large expansion. This would suggest that the initial mutation confers instability, and that the probability of increasing the length to the at-risk range is relatively small. Models of this sort have been proposed, with multiple mutation steps required for the phenotype, each with its own probability of occurrence.

Additional support for the notion of old and reasonably benign pre- (or proto) mutations comes from evidence of linkage disequilibrium among fragile X chromosomes. Using closely linked polymorphic markers, two groups have found significant differences between the haplotypes of fragile and normal X chromosomes. These may represent high-end normal alleles on a common haplotype background with an increased likelihood of increasing in size into the at-risk range. A similar mechanism has been proposed for myotonic dystrophy, where the linkage disequilibrium is quite striking (this volume, Chapter 6). Whatever the mechanism, a large number of alleles is lost with each affected generation, and these must be replenished to maintain the current frequency of the disorder. If fragile X syndrome is currently in population equilibrium, then a fairly large pool of mutant alleles would be required to maintain the disease frequency. Preliminary evidence suggests that this pool could be represented by the upper end of the normal range (>40 repeats) where unstable alleles can be found in the normal population. The alleles at this size may be as frequent as 0.5%, suggesting ~1/100 women as potential carriers of an allele at risk in future generations.

FRAGILE X DIAGNOSTICS

The advent of molecular methods for diagnosis of fragile X syndrome has diminished the role of cytogenetics in this disorder. Southern analysis remains the standard method for detection of amplified repeats, as it is simple, relatively inexpensive and, performed correctly, unambiguous. It offers the additional advantage of straightforward assay of methylation status—a feature rarely variant but critical to the disorder. PCR-based methods can be of great assistance for characterizing the sizes of pre-mutations, and for determining if normal alleles are present. However, the difficulties involved with achieving consistent amplification of the full mutation alleles (especially in the presence of normal alleles, as in carrier and affected females) remain problematic for widespread acceptance of PCR-based testing. PCR is likely to be the method of choice for population surveys, with follow-up of suspicious alleles by Southern analysis.

Is there still a role for cytogenetics in fragile X diagnostics? In view of the expense involved in cytogenetic testing, its lesser accuracy and inability to detect pre-mutation carriers, this technique cannot be justified for analysis of families with known fragile X syndrome. With new probands, especially those among the developmentally delayed population typically referred for fragile X testing, however, cytogenetic analysis is still important. The major justification for cytogenetic analysis is the possibility of other, non-fragile X, chromosomal abnormalities, which are seen at roughly the same frequency as fragile X syndrome in this population (~3%). Of lesser importance is the possible detection of one of the other fragile sites near FRAXA (E or F), as each of these fragile sites has recently been characterized at the molecular level.

FUTURE PERSPECTIVES

The identification of mutations which confer increased mutability onto themselves is a rather astounding finding which can provide insight into several previously mysterious phenomena in genetics. The idea that DNA is not necessarily inherited in the same form as in the parent or can be significantly altered from tissue to tissue within an individual is radical and calls for re-examination of some of the principles of genetics. It is a rare delight when fundamentally new phenomena are uncovered in genetics, and heritable unstable DNA represents such a new principle.

The identification of unstable trinucleotide repeats at the fragile X locus represented the first of now eight human genetic disorders with similarly unstable trinucleotide-based mutations. These are spinal and bulbar muscular atrophy, myotonic muscular dystrophy (this volume, Chapter 6), Huntington's disease (this volume, Chapter 4), spinocerebellar ataxia type 1 dentatorubral–pallidoluysian atrophy, Machado–Joseph disease and fragile X type E. All of these diseases have some features in common; in particular, each is neuro-muscular in origin, and each shows some form of variation from normal Mendelian inheritance. How many other human disorders might fit into this category? There are certainly many other genes containing trinucleotide repeats, some of which may show similar behaviour. There are also numerous genetic disorders displaying features of reduced penetrance, variable expressivity or anticipation. Trinucleotide repeat mutations may play a role in numerous familial maladies in humans.

What then might be the role of these unstable sequences? What is their normal function, and why does it go awry? It would appear that the differences between each of the above genes and diseases will increase significantly with more study. Already, the positions and behaviours of the repeats are sufficiently different that it is unlikely that each serves an identical function. Evolutionary conservation may provide a clue to function; however, such mutations have only been demonstrated in the human thus far, despite the presence of the repeats in other model organisms such as the mouse. Does this behaviour reflect a fundamental difference in the ability of the human DNA replication machinery to handle GC-rich trinucleotide repeats? The questions are endless, but the answers will be forthcoming.

REFERENCES

Abitol M, Menini C, Delezoide A-L et al. (1993) Nucleus basalis magnocellularis and hippocampus are the major sites of FMR-1 expression in the human fetal brain. Nature Genet., 4, 147–153.

Bell MV, Hirst MC, Nakahori Y et al. (1991) Physical mapping across the fragile X: hypermethylation and clinical expression of the fragile X syndrome. Cell, 64, 861–866.

Fu YH, Kuhl DPA, Pizzuti A et al. (1991) Variation of CGG repeat at the fragile X site results in genetic instability: resolution of the Sherman paradox. Cell, 67, 1047–1058.

Heitz D, Rousseau F, Devys D et al. (1991) Isolation of sequences that span the fragile X and identification of a fragile X-related CpG island. Science, 251, 1236–1239.

Hinds HL, Ashley CT, Sutcliffe JS et al. (1993) Tissue specific expression of FMR-1 provides evidence for a functional role in fragile X syndrome. Nature Genet., 3, 36–43.

Kremer EJ, Pritchard M, Lynch M et al. (1991) Mapping of DNA instability at the fragile X to a trinucleotide repeat sequence p(CCG)n. Science, 252, 1711–1714.

Laird CD (1987) Proposed mechanism of inheritance and expression of the human fragile-X syndrome of mental retardation. Genetics, 117, 587–599.

Lubs HA (1969) A marker X chromosome. Am. J. Hum. Genet., 21, 231–244.

Martin JP and Bell J (1943) A pedigree of mental defect showing sex-linkage. J. Neurol. Psychiatry, 6, 154–157.

Oberlé I, Rousseau F, Heitz D et al. (1991) Instability of a 550-base pair DNA segment and abnormal methylation in fragile X syndrome. Science, 252, 1097–1102.

Pieretti M, Zhang F, Fu YH et al. (1991) Absence of expression of the FMR-1 gene in fragile X syndrome. Cell, 66, 817–822.

Reyniers E, Vits L, De Boulle K et al. (1993) The full mutation in the FMR-1 gene of male fragile X patients is absent in their sperm. Nature Genet., 4, 143–146.

Sherman SL, Morton NE, Jacobs PA and Turner G (1984) The marker (X) syndrome: a cytogenetic and genetic analysis. Ann. Hum. Genet., 48, 21–37.

Verheij C, Bakker CE, de Graaff E et al. (1993) Characterization and localization of the FMR-1 gene product. Nature, 363, 722–724.

Verkerk AJMH, Pieretti M, Sutcliffe JS et al. (1991) Identification of a gene (FMR-1) containing a CGG repeat coincident with a breakpoint cluster region exhibiting length variation in fragile X syndrome. Cell, 65, 905–914.

Vincent A, Heitz D, Petit C et al. (1991) Abnormal pattern detected in fragile-X patients by pulsed-field gel electrophoresis. Nature, 349, 624–626.

Warren ST, Knight SJL, Peters JF et al. (1990) Isolation of the human chromosomal band Xq28 within somatic cell hybrids by fragile X site cleavage. Proc. Natl Acad. Sci. USA, 87, 3856–3860.

Wöhrle D, Hennig I, Vogel W and Steinbach P (1993) Mitotic stability of fragile X mutations in differentiated cells indicates early post-conceptual trinucleotide repeat expansion. Nature Genet., 4, 140–142.

Yu S, Pritchard M, Kremer E et al. (1991) Fragile X genotype characterized by an unstable region of DNA. Science, 252, 1179–1181.

FURTHER READING

Burke DT, Carle GF and Olson MV (1987) Cloning of large segments of exogenous DNA into yeast by means of artificial chromosome vectors. Science, 236, 806–812.

Hagerman RJ (1991) Physical and behavioral phenotype. In Fragile X Syndrome: Diagnosis, Treatment and Research, eds Hagerman R and Silverman A (pp. 3–68). Baltimore: Johns Hopkins University Press.

Laird CD, Lamb MM, Sved J and Thorne J (1991) Modeling the inheritance and expression of fragile X syndrome, with emphasis on the X-inactivation imprinting model. In Fragile X Syndrome: Diagnosis, Treatment and Research, eds Hagerman R and Silverman A (pp. 228–251). Baltimore: Johns Hopkins University Press.

Nussbaum RL and Ledbetter DH (1989) The fragile X syndrome. In The Metabolic Basis of Inherited Disease, 6th edn, eds Scriver CR, Beaudet AL, Sly WS and Valle D (pp. 327–341). New York: McGraw-Hill.

Oberlé I, Drayna D, Camerino G, White R and Mandel J-L (1985) The telomeric region of the human X chromosome long arm: presence of a highly polymorphic DNA marker and analysis of recombination frequency. Proc. Natl Acad. Sci. USA, 82, 2824–2828.

Oostra BA and Verkerk AJMH (1992) Review. The fragile X syndrome: isolation of the FMR-1 gene and characterization of the fragile X mutation. Chromosoma, 101, 381–387.

Oostra BA, Jacky PB, Brown WT and Rousseau F (1992) Guidelines for the Diagnosis of Fragile X syndrome. International Fragile X Conference, Aspen, Colorado. Dillon, CO: Spectra.

Pennington BF, O'Connor RA and Sudhalter V (1991) Toward a neuropsychology of fragile X syndrome. In Fragile X Syndrome: Diagnosis, Treatment and Research, eds Hagerman R and Silverman A (pp. 173–201). Baltimore: Johns Hopkins University Press.

Prouty LA, Rogers RC, Stevenson RE et al. (1988) Fragile X syndrome: growth, development, and intellectual function. *Am. J. Hum. Genet.*, **30**, 123–142.

Sutherland GR (1977) Fragile sites on human chromosomes: demonstration of their dependence on the type of tissue culture medium. *Science*, **197**, 265–266.

Sutherland GR, Hecht F, Mulley JC, Glover TW and Hecht BK (1985) The fragile X: intelligence, behaviour and treatment. In *Fragile Sites on Human Chromosomes* (pp. 113–133). New York: Oxford University Press.

Turner G, Opitz JM, Brown WT *et al.* (1986) Conference report: Second International Workshop on the fragile X and X-linked mental retardation. *Am. J. Hum. Genet.*, **23**, 11–67.

Warren ST, Zhang F, Licameli GR and Peters JF (1987) The fragile X site in somatic cell hybrids: an approach for molecular cloning of fragile sites. *Science*, **237**, 420–423.

6 The Molecular Genetics of Myotonic Dystrophy

J. DAVID BROOK and HELEN G. HARLEY

THE MYOTONIC DYSTROPHY PHENOTYPE

MYOTONIC DYSTROPHY IS A GENETIC DISORDER

Descriptions of myotonic dystrophy (DM; also known as dystrophia myotonica or myotonia atrophica) appeared in medical journals as early as the turn of the century. These publications detailed individuals with myotonia, and progressive muscle weakness with a distinct distribution of muscle involvement. Although we now consider DM to be a multisystem disorder, this description of DM still provides an accurate summation of its main features.

These early studies also provided clear evidence that DM is a genetic disorder. By studying family trees of individuals with DM it could be seen that many family members over several generations were affected. Careful analysis of these pedigrees showed that the disorder affected and was transmitted by both males and females (therefore the gene was unlikely to be found on either of the sex-determining chromosomes, the X or the Y) and that one altered copy of the DM gene was sufficient to cause the specific muscle changes. Thus DM was classified as an autosomal dominant disorder.

There was also strong evidence from linkage studies that, despite great variation in the clinical phenotype both between and within families, DM was a genetically homogeneous disorder, i.e. that all cases were due to a mutation in the same gene. Some early studies also suggested that the DM gene was incompletely penetrant and that some gene carriers remain asymptomatic throughout their life. Minimally affected individuals are often asymptomatic well into old age, although detection of mild signs of the disorder will depend on how thoroughly and at what age a patient has been examined. This difficulty of diagnosis of minimal cases could explain the observation of incomplete penetrance. However, it is now apparent that the nature of the mutation underlying the DM phenotype explains why some individuals are asymptomatic.

DM is the most frequent of the adult muscular dystrophies, with an

Molecular Genetics of Human Inherited Disease. Edited by D.J. Shaw
Published 1995 by John Wiley & Sons Ltd

estimated prevalence of 1 : 8000. This is likely to be an underestimate because of the nature of the mutation causing the DM phenotype; the reasons for this will become apparent when discussed under 'Phenotype–genotype correlations and clinical applications'. This estimate of prevalence is a global one, with some populations apparently lacking the DM mutation (e.g., the native population of the African subcontinent), while a few isolated populations have a much higher prevalence due to a founder effect by one or a few gene carriers (e.g., northern Quebec, where the incidence is an estimated 1 per 530 births). Reports exist of mating between two DM gene carriers but it was not until 1994 that a proven case of a homozygous DM individual was identified. The individual was the result of a consanguinous mating between a mildly affected father and his daughter who had typical adult onset. The male offspring of this mating was congenitally affected with a very severe phenotype and rapid progression of disease. These clinical observations suggested that homozygosity results in a more severe form of disease.

The rate of spontaneous abortion and neonatal death rate are sharply increased in the pregnancies of women with myotonic dystrophy. Total reproductive loss has been estimated to be 30% (17% spontaneous abortion, 11% neonatal death, 2% stillbirth). Reproductive fitness also decreases with the severity of disease, particularly in males. It might therefore be expected that carriers of the DM mutation should decrease in frequency, but this does not appear to be the case, as families with no previous history of DM are still being identified. Possible mechanisms for this are discussed under 'Phenotype–genotype correlations and clinical applications'.

INTER- AND INTRAFAMILIAL VARIATION OF THE MYOTONIC PHENOTYPE

The main features of DM are myotonia, and progressive weakness and wasting of muscle with any combination of cardiac conduction defects, smooth muscle involvement, mental changes, hypersomnia, ocular cataract, testicular atrophy and endocrine dysfunction. If the disease is manifest at birth there may be neonatal hypotonia with decreased electrophysiological activity, facial diplegia, severe respiratory distress and mental retardation. For ease of clinical diagnosis, patients can be placed in one of three graded categories:

(1) Minimal—cataract is the main clinical feature and neuromuscular abnor-malities are absent or mild, with onset of symptoms occuring in middle or old age.
(2) Adult onset—myotonia and progressive muscle weakness are clearly present, with onset in adolescence or early adult life.
(3) Congenital—symptoms are present at birth or *in utero*. These include respi-ratory insufficiency, hypotonia and developmental delay in survivors.

Mild disease in one family member does not preclude severe disease in another, with the exception of congenitally affected individuals where a high proportion of siblings also show this severe form. All three forms of the disorder are commonly found within a single family.

CLINICAL DIAGNOSIS OF MYOTONIC DYSTROPHY

The multisystemic nature of DM, i.e. many different tissues in the body are affected, is such that the abnormalities detected in both muscle and other systems are of a specific nature. The clinical picture is quite complex (for an extensive discussion see Harper, 1989) but the main features will be dealt with briefly here.

Cataract was noted as a consistent feature of minimal DM even in the earliest studies. In particular, different branches of family trees could be linked by relatives who had cataract but no overt symptoms of the disorder. During the early stages of development, the cataracts can be identified as DM-specific multicolour subcapsular opacities. However, they are not always apparent in the early stages of DM and, once they have matured, they are indistinguishable from any other type of cataract, although they will tend to occur bilaterally rather than unilaterally. These specific lens changes are perhaps the most consistent extramuscular feature of DM. However, some patients with severe muscle problems may have insignificant opacities while others have mature opacities when any muscle symptoms are minimal or totally absent. Cataracts are not common in infants with congenital DM, but they may develop in later life.

Adult-onset DM is characterized by myotonia (impaired muscle relaxation resulting from repetitive discharge arising in the muscle itself) and progressive muscle weakness and wasting. Myotonia of the grip or percussion myotonia of the thenar eminence can be elicited in almost every symptomatic patient, being most noticeable in those with relatively minor muscle weakness and wasting. A firm handshake will often detect whether or not active myotonia is present: myotonia will result in a slow release of the hand grip due to a delayed ability to relax the hand muscles. In some cases percussion myotonia—the involuntary contraction of a muscle when it has been struck with a tendon hammer—will be detectable when grip myotonia is not apparent due either to the mildness of the myotonia or to the patient's ability to disguise grip myotonia by altering the handshake. Myotonia may be interpreted by the patient as a form of muscle stiffness, although many seem unaware of it, perhaps because its effect is diminished by repeated contraction and relaxation of the muscles, and by elevated temperature. The distinctive pattern of muscle involvement allows clinicians specifically to identify those individuals with DM as opposed to one of the other adult muscle disorders (Table 6.1). There is a generalized facial weakness and ptosis (droopy eyelids) such that there is a striking lack of facial expression in

Table 6.1. Muscle involvement in DM and other adult myopathies

	DM	Fascio-scapulo-humeral dystrophy	Autosomal recessive limb girdle dystrophy	Becker dystrophy
Facial weakness	++	++	+	−
Ptosis	++	++	−	−
Jaw muscles	++	+	−	−
Sternomastoids	++	+	−	−
Shoulder girdle	±	++	+	+
Pelvic girdle	±	+	++	++
Proximal limb muscles	+	+	++	++
Distal limb muscles	++	±	+	+
Myotonia	++	−	−	−
Pseudo-hypertrophy	±	−	±	++

affected individuals. This is clearly evident in individuals with adult onset and in congenital DM, but may be absent in mildly affected cases (see Fig. 6.1). There is also some involvement of distal limb muscles, with more general limb weakness as the disease progresses. Severe weakness of weight-bearing muscles occurs, if at all, at a later stage of the disease. This means that limited mobility is retained even at advanced stages of general muscle weakness.

Abnormalities in mental function are also noted at an early stage of the disease. Apart from the mental retardation associated with congenital cases, this aspect of the phenotype is harder to quantify. In adult-onset cases, the following observations have been made: reduced initiative, inactivity, apathy, and social deterioration over several generations. Documentation of these aspects is often anecdotal, but there is general agreement that the degree of general apathy and inertia is not readily explicable by the degree of neuromuscular disability.

The clinical differentiation of DM from other progressive disorders is rarely difficult, especially when the family as a whole is studied. The systemic features of adult-onset DM, listed in Table 6.2, are also an important aid to differential diagnosis from other inherited muscle diseases (myopathies), which fall into three categories:

(1) The muscular dystrophies characterized by progressive muscle degeneration, with wasting and weakness.
(2) The myotonic syndromes, and several other disorders that result in muscle stiffness, clinically resembling myotonia.
(3) Non-progressive myopathies, in particular those with congenital onset. A detailed discussion of these can be found in Harper (1989).

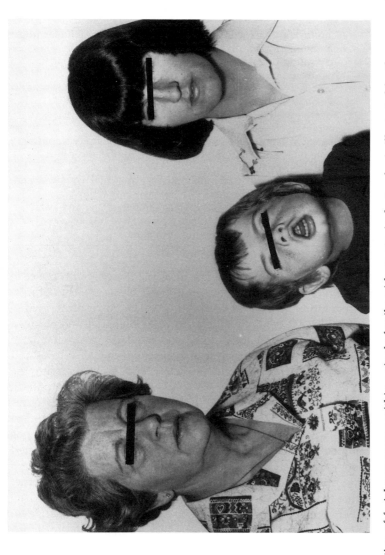

Fig. 6.1. Variation of facial appearance within a single family with myotonic dystrophy is illustrated by three generations of gene carriers: (i) on the left, the grandmother had bilateral cataracts removed at the age of 40 years, has percussion myotonia but no other muscle symptoms and no facial weakness; (ii) on the right, her daughter had no obvious symptoms until after the birth of her severely affected child, but examination showed that she has moderate facial weakness and ptosis, with active and percussion myotonia; lens opacities were noted at her initial examination and cataract extraction has subsequently been performed; (iii) her child had congenital myotonic dystrophy, and shows marked facial diplegia and jaw weakness, present from birth

Table 6.2. Systemic involvement in myotonic dystrophy

System	Principal involvement
Smooth muscles	Gastrointestinal tract, in particular the oesophagus (problems with swallowing) and colon; uterus; other sites may also be affected
Heart	Conduction defects, in particular heart block, arrhythmias (most commonly in the atria); and, rarely, cardiomyopathy
Lungs	Aspiration pneumonia from oesophageal involvement, hypoventilation, hypersomnia
Peripheral nerve	Variable involvement, rarely clinically significant; minor sensory loss may occur
CNS	Severe involvement in the congenital form; variable loss of mental functions; hypersomnia
Endocrine	Testicular tabular atrophy; diabetes (rarely clinically significant); sometimes abnormalities of growth hormone and other pituitary functions
Eye	Cataract, retinal degeneration, ptosis, extraocular weakness and, rarely, ocular hypotonia
Skeletal	Cranial hyperostosis, air sinus enlargement; jaw and palate involvement; talipes (childhood cases); and, rarely, scoliosis
Skin	Premature balding; calcifying epithelioma

Congenital DM is a highly characteristic form of the disease with onset *in utero*. The major clinical features are bilateral facial weakness, hypotonia, delayed motor development, mental retardation, neonatal respiratory distress, feeding difficulties, talipes, hydramnios in later pregnancy and reduced fetal movement. It is rare for all features to be documented in a single infant. In older children with congenital onset (Fig. 6.1), immobile facial features and an open jaw due to weakness of the muscles tend to exaggerate the impression of mental dullness, although many are mentally impaired. Myotonia is not a feature of congenital DM at birth, although EMG studies may detect the presence of myotonic potentials in the first year of life. Clinical myotonia is not usually detected in children under 2 years of age, but has invariably developed by the end of their tenth year. Congenitally affected infants are almost exclusively born to symptomatic mothers, suggesting a possible maternal effect in its aetiology. It was thought that the maternal effect could be due to maternal transmission of defects in the mitochondrial genome. However, there does not appear to be a significant change in mitochondrial size or number, or any abnormality in mitochondrial DNA gene carriers as compared to controls. Although most severely affected infants are born with hypotonia, this may disappear within weeks, which may be indirect evidence for a maternal environmental effect on the fetal phenotype. Evidence for an effect of the maternal phenotype on that of the fetus is discussed under 'Phenotype–genotype correlations and clinical applications'.

MUSCLE PATHOLOGY AND STRUCTURAL CHANGES IN PERIPHERAL NERVE AND THE EYE

Muscle biopsies taken from a variety of muscle groups can be examined using a wide range of histological, histochemical and ultrastructural techinques. While there is no single change diagnostic of DM, the overall pattern of abnormalities seen is distinctive. The characteristic features are increased nuclei in the muscle fibres, these nuclei often occuring in chains; ringed fibers, with sarcoplasmic masses (homogeneous areas of disorganized intermyofibrillary material) often observed nearby; atrophy of type 1 fibres; and an increased number of fibres in the muscle spindles. Blood vessels and the basement membrane are normal, as is the innervation apart from terminal innervation often showing increased arborization of the nerve fibres. The increased number of fibres in the muscle spindles are caused by splitting and refusion of the fibres, with motor end plates dispersed over most of the fibre. It is not known whether this increase in number is caused by arrested development of the spindle or from the mechanical stresses induced by myotonia.

In general, any changes in peripheral nerve are minor and of little clinical importance. They do not seem to be involved in the pathogenesis of either myotonia or muscle degeneration. The principal abnormality is the terminal arborations, which are usually extensive and elongated. These changes may also be related to the mechanical stress of myotonia and to changes in the muscle spindle.

The changes seen in muscle of congenitally affected infants show a number of unusual features, although the more typical changes are usually seen as the infants grow older. There is a striking hypoplasia of muscle which is especially evident in respiratory muscle. Muscle fibres appear reduced in diameter and number and round in cross-section. The central position of the nuclei in the fibres may reflect immaturity of the fibre rather than a secondary inward migration of the nuclei. When looked at histologically, there is a characteristic lack of oxidative enzyme activity at the periphery of the type 1 fibres and a failure of the fibres to differentiate into specific types. Active degenerative changes such as fibre splitting and alterations in the muscle spindle are largely absent in infancy. The prominence of satellite cells in congenital DM, in contrast to most other congenital myopathies, further supports the concept that there has been an arrested development of muscle rather than active degeneration.

Biochemical and structural changes in the lens have not been much investigated in DM patients. In one patient it was found that the potassium concentration was greatly reduced from 635 mg per 100 ml in the normal lens and 92–309 mg/100 ml in senile cataracts to 10.9 mg/100 ml in myotonic cataracts. The elevated serum γ-glutamyl transpeptidase observed in DM could predispose to cataract.

WHY STUDY FAMILIES WITH MYOTONIC DYSTROPHY?

There were several features which made DM a challenging disease to study. First is the great variation in the degree of severity of the disease both within and between families. This can range from fully asymptomatic gene carriers to individuals affected with severe muscle impairment and mental retardation from birth. Secondly, this latter clinical group—the congenitally affected infants—are born almost exclusively to women who are symptomatic gene carriers, suggesting that a maternal factor may be involved in producing this very severe phenotype. Thirdly, there appeared to be many families in which the severity of symptoms increased and the age at which they were first noticed (age at onset) decreased from one generation to the next. This phenomenon, called anticipation, was the source of much discussion between people studying DM. Some believed that it was a real biological effect. Others thought that it was seen only because families were first identified when a severely affected individual was born to the family and that other family members were therefore likely to be less severely affected. This is referred to in the literature as ascertainment bias. As we shall see below, under 'Clinical applications', anticipation can now be explained in molecular terms. Finally, there was no proven case of new mutation. Most individuals with the DM phenotype could be traced back to a family with a history of the disease although it was not always possible, for example in a three-generation pedigree, to identify which of the grandparents was the gene carrier as both appeared asymptomatic. There were also many examples of apparently healthy parents producing more than one affected child.

The detailed clinical studies summarized above, combined with extensive family collection by a number of clinical investigators, provided the starting material for a positional cloning approach to identify the DM gene. This is described in the following section.

CLONING THE MYOTONIC DYSTROPHY GENE

LINKAGE ANALYSIS

In order to locate and clone the DM gene, a series of techniques were employed which are collectively referred to as positional cloning. Essentially there are two elements to positional cloning: linkage analysis and physical mapping. The first step is to identify genetic linkage between a polymorphic marker and the disease gene of interest, and to assign a chromosomal location to the disease gene. Today the polymorphic markers of choice to search for a disease locus are the microsatellite DNA markers, also known as CA repeats. The availability of around 1000 such markers, with high degrees of heterozygosity, distributed more or less evenly around the genome represents a very powerful resource. The chromosomal location of these markers and their order along the chromosomes have been established. Microsatellite DNA markers can be rapidly typed in polymerase chain reaction (PCR) assays which allows a high throughput of samples. Thus for an autosomal dominant trait with half a dozen well-characterized three-generation families involving 50 or so individuals of whom roughly half are affected, it is likely that linkage to a microsatellite DNA marker can be demonstrated within 1 year by a single laboratory worker. The development of microsatellite DNA markers has occurred since 1991, hence such a powerful resource was not available when linkage to DM was originally being sought. Thus much of the linkage analysis for DM was performed using the more time-consuming procedure of radiolabelled DNA probe hybridization to Southern blots to follow the segregation of restriction fragment length polymorphism (RFLPs) in DM families.

The earliest linkage studies on DM, however, preceded the development of DNA technology and RFLPs and were thus seriously limited by the lack of available polymorphic loci. Red cell antigens, isoenzymes and human leucocyte antigen (HLA) constituted the best polymorphic markers available. Despite the dearth of polymorphic markers, it is intriguing that the first autosomal linkage group discovered in man involved markers now known to be syntenic with DM. Indeed, when linkage was discovered between the lutheran blood group (Lu) and what was later shown to be the secretor (Se) locus in the early 1950s, it was suggested that DM might form part of the same linkage group. It was not until almost 20 years later in the early 1970s that this suggestion was confirmed. It took a further 10 years for this linkage group to be assigned to chromosome 19, this being achieved because of two separate pieces of work involving the complement component gene C3. First, by virtue of protein polymorphism, the C3 gene was shown to form part of the Se-Lu-Le linkage group. Shortly thereafter the C3 gene was cloned and assigned to chromosome 19 using DNA techniques to localize a gene probe against a panel of somatic cell hybrid DNAs. Thus the search for the DM gene using recombinant DNA techniques had begun. A comparison of chromosome 19 linkage maps from 1985 and 1994 is shown in Fig. 6.2. The 1985 map, one of the first of its kind, is based on blood group markers and RFLPs. The 1994 map is almost entirely based on microsatellite markers.

DEFINING THE INTERVAL CONTAINING MYOTONIC DYSTROPHY

The second stage of a positional cloning strategy to identify and clone a human disease gene is the narrowing down of the interval containing that gene and the identification of flanking markers. During the 1980s, prior to the discovery of microsatellite DNA markers, this was achieved by analysing the segregation of RFLPs in DM families. As mentioned earlier this was both laborious and time-consuming; nevertheless it represented at that time the most effective method for 'homing in' on the DM gene.

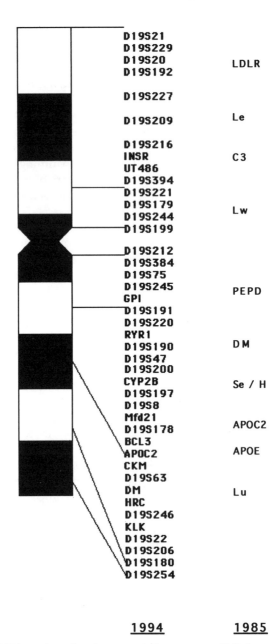

| | 1994 | 1985 |

Fig. 6.2. The 1994 version of an integrated chromosome 19 linkage map published by Buetow *et al.* (*Nature Genetics*, 1994, **6**, 391–393), where the distances between markers has been established. This compares with the first genetic map of chromosome 19 based on two-point LOD tables published in 1985 by Sherman *et al.* (*Cytogenetics and Cell Genetics*, 1985, **40**, 742) in which the order of markers and distances between them are much less certain. From Sherman *et al.* 1985, with permission

The first RFLP shown to be linked to DM was one associated with the C3 gene (Davies *et al.*, 1983). In this case a significant result was obtained in DM males. However, in DM females no linkage was demonstrated, reflecting sex-specific differences in recombination frequency that are often seen in genetic maps. The usefulness of C3 as a marker co-segregating with DM was somewhat limited for two reasons. First, the polymorphism frequency was low. Second, the distance between DM and C3 was not particularly close. In fact it was rather fortunate that linkage was ever demonstrated between C3 and DM even in males, as the most recent genetic data show that C3 and DM are at opposite ends of different arms of chromosome 19.

Two years after linkage between C3 and DM was found, a marker that showed close linkage to DM was identified. Like many advances in this field the development of a closely linked marker for DM was fortuitous. The genes encoding three different apolipoproteins—C1, CII and E—were cloned and localized to chromosome 19. Polymorphisms were identified with these markers and a comprehensive study of DM families showed that APOC2 is closely linked to DM. The maximum LOD score of 7.8 was at a recombination fraction of 4%. This represented the first clinically useful DNA marker for prenatal diagnosis of DM (Shaw *et al.*, 1985).

SOMATIC CELL HYBRID STUDIES

Once the approximate location of the DM gene had been identified through segregation analysis with a closely linked marker, stage two of the positional cloning strategy, which is physical mapping, could be employed. The foundation for much of this work was laid during the mid-1960s when selective systems for somatic cell genetic studies were developed, which, when combined with techniques for cell fusion, led to the construction of rodent–human somatic cell hybrids. Studies on rodent–human somatic cell hybrid lines revealed a preferential loss of human chromosomes from such cell lines. Deficiency in the genes for hypoxanthine phosphoribosyl transferase (HPRT) and thymidine kinase (TK) in rodent mutant cell lines can be complemented by homologous genes on human chromosomes X and 17 respectively. Thus these chromosomes can be selectively retained in hybrid cell lines. If donor fibroblast lines which carry translocation chromosomes between X and 17 and another chromosome, for example chromosome 19, are used in fusion experiments it is possible to retain the portion of the other chromosome which is joined on to the part of the X or 17 that is selectively retained. In this way, using donor fibroblasts containing translocations involving chromosome 17 or X and 19, it was possible to develop a panel of hybrid cell lines containing different pieces of chromosome 19. In the absence of other material from chromosome 19 it is possible to localize DNA markers to various regions of the chromosome. Thus a series of hybrid cell lines were developed for this purpose (Fig. 6.3). The early chromosome 19 localization

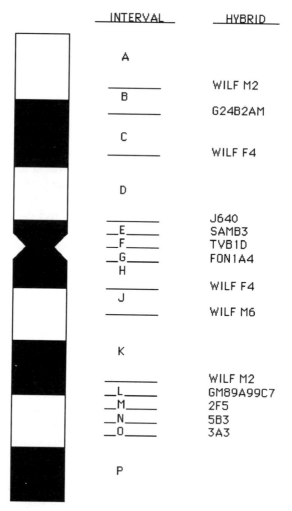

Fig. 6.3. The various regions of chromosome 19 to which markers can be localized based on available somatic cell hybrid lines

data indicated that the C3 gene maps to the distal end of the short arm. Shortly thereafter peptidase D, in addition to other markers forming part of the DM linkage group, was localized to the central part of the chromosome, hence the linkage group could be orientated. Further studies involving hybrid cell lines and others using *in situ* hybridization provided strong evidence to suggest that the markers most closely linked to DM—the apolipoprotein genes APOE, APOC1 and APOC2—localize to band 19q13.2 (see Fig. 6.2). Thus the search for the DM gene was focused in this interval.

HOMING IN ON THE DM GENE

The progress towards identification and cloning of the DM gene and other inherited disease genes can be directly related to the development of techniques for analysing and cloning large DNA fragments. An example of this technological advance is pulsed-field gel electrophoresis. DNA fragments larger than 30–50 kb cannot be resolved by conventional agarose gel electrophoresis. By switching the current at perpendicular angles every 1–2 min, it was found that DNA fragments of greater than 2 million base pairs could be resolved on agarose gels. Using restriction enzymes which cut infrequently in the genome such as Not1 and Mlu1 it is possible to generate DNA fragments in this size range, and construct long-range restriction maps. Such maps were constructed using the most closely linked markers to DM, producing a comprehensive long-range map of the interval containing DM.

The fortuitous localization of two other genes to the DM region of chromosome 19 was to play a significant part in the progress toward gene identification. The genes are creatine kinase muscle form (CKM) and the excision repair gene, ERCC1. Multipoint linkage analysis indicated that CKM maps between APOC2 and DM, hence making it the most closely linked marker to DM at that time. Furthermore, studies on the DNA repair deficient rodent cell line UV20 revealed that the DNA repair mutation could be complemented by the human gene ERCC1, which maps very close to CKM on chromosome 19. This placed ERCC1 very close to DM also. This synteny was exploited by selecting for the retention of ERCC1 in DNA repair deficient rodent cell lines fused to human cell lines. It was possible to develop hybrids containing a single copy of chromosome 19 as their only human material, as well as other hybrids containing fragments of chromosome 19. The development of these hybrid cell lines was a crucial step on the path to the identification of the DM gene. Essentially two different approaches were adopted to identify the DM gene, both of which utilized these cell lines. One approach required the construction of further somatic cell hybrids containing even smaller amounts of human chromosome 19 and the other involved a cosmid walk using a library constructed from DNA of the chromosome 19 only somatic cell hybrid.

In the first approach radiation-reduced hybrids were made (Brook et al., 1992a). This procedure is summarized in Fig. 6.4. A donor cell line, 20XP, carrying a fragment of chromosome 19 containing the DM region in addition to a few pieces of other human chromosomes, was irradiated with a lethal dose of X-rays. The X-irradiation kills the donor cell line and fragments its chromosomes. The X-ray dosage determines the exent of the fragmentation. Typically 5000–8000 rads are employed in this process. The irradiated donor cell line was then fused to the DNA repair-deficient rodent cell line UV20 and selection for the retention of the human ERCC1 gene was introduced. Most of the DNA fragments are lost in the resulting hybrids, except for the selectable

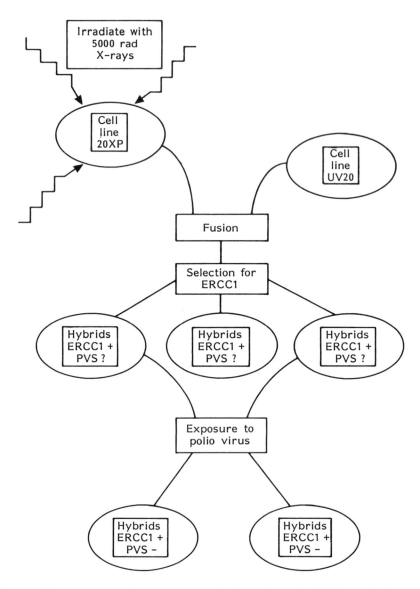

Fig. 6.4. The strategy used to construct radiation-reduced hybrids containing small amounts of DNA (2–3 megabases) around the DM locus

marker ERCC1 and the immediate flanking DNA. The gene encoding the polio virus receptor (PVR) had previously been shown to map in the same interval as ERCC1 and CKM but on the opposite side of ERCC1 to the DM gene. This receptor could therefore be used as a negative selectable marker to kill cells containing the PVR gene. Cells were exposed to polio virus whilst maintaining selection for ERCC1. In this way hybrids which retained a minimal amount of chromosome 19 from the proximal side of ERCC1 (i.e., the opposite side to that containing DM) could be selected. Thus a hybrid cell line was produced which contained only 2–3 Mb of human DNA from chromosome 19 which included those markers most closely associated with DM. DNA from this cell line was used to construct a genomic phage library from which a series of markers were derived to saturate and 'clone out' the DM-containing interval. Further radiation-reduced hybrids and detailed pulsed-field gel maps were required to position clones with respect to each other.

A different approach to cloning the DM gene also relied on the construction of a chromosome 19 only human hamster hybrid cell line. In this case the hybrid provided a source of DNA for the construction of a cosmid library to allow high-density coverage of clones along chromosome 19. A chromosome walk was initiated from CKM distally towards the DM region. In order to perform such a walk cosmids were identified which cross-hybridize with the starting point—CKM. The ends of these cosmids were used as probes to rescreen the library and identify the next cosmid. This procedure was repeated many times until a series of overlapping clones was established (Shutler *et al.*, 1992).

IDENTIFICATION OF ENLARGED DNA FRAGMENTS IN DM PATIENTS

Given the intensity of the search by several different research groups throughout the world, it was not surprising that the myotonic dystrophy gene and the mutation underlying the disease were eventually discovered. What was surprising, however, was the nature of the mutation causing myotonic dystrophy. The saturation coverage of the DM interval by DNA markers led to the discovery of an enlarged DNA fragment in DM patients. Papers from three groups reported the finding that specific DNA probes identified a polymorphic EcoR1 restriction fragment of 9 kb or 10 kb in normal individuals. In DM patients, however, the same probes identify additional restriction fragments greater in size than 10 kb which in some instances, particularly the larger ones, appear as smears (Harley *et al.*, 1992). It was also apparent from this experiment that the larger restriction fragments were unstable, such that they showed further enlargement each time they were transmitted from parent to offspring. Enlarged DM DNA fragments are shown in Fig. 6.5.

Three papers published shortly afterwards documented that this enlarged DNA fragment was due to the expansion of a repeated trinucleotide DNA sequence (CTG). The length of this repeat is variable in the normal population, where the number of copies ranges from 5 to 37. In DM patients, however, this repeat is greatly expanded, and the extent to which it is expanded correlates with the severity and increasingly early age-at-onset of

132

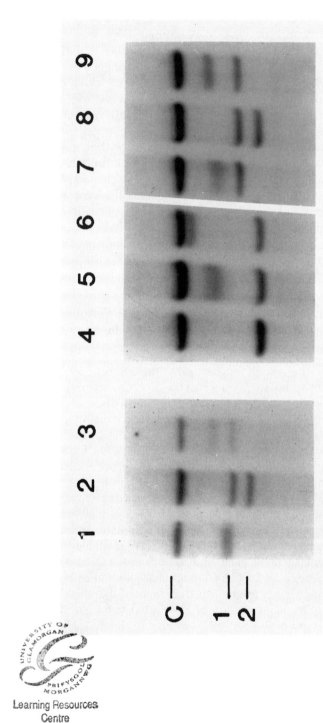

Fig. 6.5. Expanded alleles at the DM locus. C is a constant band present in both patients and unaffected individuals. Bands 1 and 2 are different alleles of an RFLP which is located within an intron of the DMPK gene. The smeared bands between band 1 and the constant band are expanded alleles in DM patients (lanes 1, 3, 5, 6, 7 and 9). Unaffected individuals are shown in lanes 2, 4 and 8

the disease (Brook *et al.*, 1992b). Thus minimally affected DM patients have 50–100 copies of the repeat, whereas the most severely affected congenital DM patients may have several thousand copies of this triplet. The sequence of an expanded repeat from a minimally affected patient with 50 repeats is shown in Fig. 6.6. The finding that an expanded trinucleotide DNA sequence was responsible for DM is remarkable, particularly in view of the previously reported discovery that fragile X mental retardation syndrome (FRAXA), which shares many of the unusual genetic features of DM, was caused by a similar mechanism. In the case of FRAXA the expanding triplet is CGG and the repeat is associated with the 5′ end of a gene, FMR1.

Analysis of the DNA flanking the DM-associated triplet repeat revealed cross-species sequence conservation, indicative of coding sequences. cDNA library screening was performed with this DNA and several clones were identified. Characterization of these clones revealed that the cDNAs share sequence similarity with members of the protein kinase gene family. The strongest similarity is with the cAMP-dependent serine/threonine kinases. Furthermore, analysis of the sequence of DMPK (myotonic dystrophy protein kinase) revealed that the triplet repeat is located in the 3′ untranslated region (UTR); that is, the triplet repeat does not encode protein.

Thus, as for many other inherited disorders, the identification of the underlying basis of DM represented a triumph for positional cloning techniques. Over the course of 10 years it was possible to move from DM chromosomal location to cDNA cloning. The next stage in this process is to unravel the mechanism by which this expanded triplet may generate the DM phenotype. The most recent developments in this area will be considered under 'The molecular basis of myotonic dystrophy'. In the next section we consider the clinincal implications and applications of this finding.

PHENOTYPE–GENOTYPE CORRELATIONS AND CLINICAL APPLICATIONS

The discovery that expansion of an unstable CTG repeat is the mutational basis underlying DM has provided an explanation for many of the puzzling clinical and genetic aspects of the disorder. It is now possible to explain the marked variability in severity and age at onset, both between families and also between different generations of the same family, by the nature of a specific, and newly described, type of molecular change in the DNA of affected individuals—that of an unstable or dynamic mutation.

THE RELATIONSHIP OF SIZE OF REPEAT TO PHENOTYPE IN INDIVIDUAL PATIENTS

The DM phenotype is caused by a change in the number of CTG repeats in the 3′-UTR of the DMPK gene on human chromosome 19q13.3. This repeat is

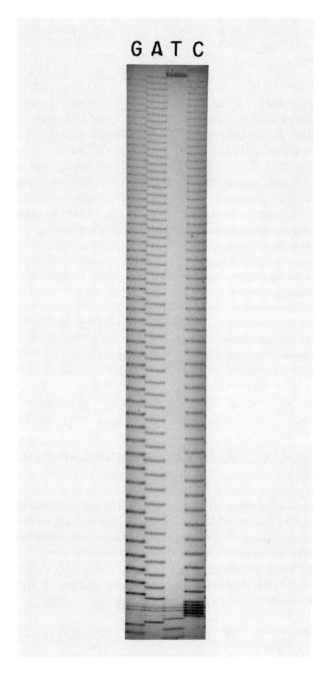

Fig. 6.6. The DNA sequences of an expanded triplet repeat from a minimally affected DM patient. Published with permission of CELL Press

polymorphic, i.e. there is a range of different repeat numbers in a control population of individuals not affected by DM. The extent of this variation is shown in Fig. 6.7. The number of CTG repeats detected in controls varies between 5 and 37 repeats, and shows a bimodal pattern when represented graphically, with alleles of 5 and 11–13 being the most common in the Caucasian population. Each individual has a high probability of being heterozygous or carrying different-sized CTG repeats on their two chromosome 19s, making this a useful polymorphism for studying the common inheritance of this region of chromosome 19 in apparently unrelated DM families. Individuals known to be affected with DM have one chromosome 19 with a CTG repeat number within the range seen in the control population (5–37) and their other chromosome 19 has a repeat number of from about 50 up to several thousand CTGs. This is often referred to as the expanded allele because such a large increase in the number of repeats makes this part of chromosome 19 much bigger in size than its normal partner. Although variable in size, the expanded allele still shows simple Mendelian inheritance in DM families.

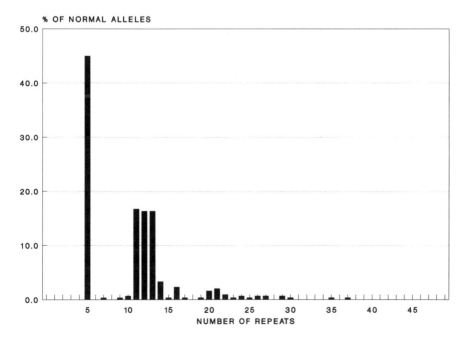

Fig. 6.7. Distribution of CTG alleles within a control population. The x axis represents the number of CTG repeats on a chromosome; the y axis represents the proportion of alleles observed with this repeat length. A total of 328 chromosomes were analysed

When affected individuals are classified according to the phenotypic criteria described in the first section and the size of their expanded alleles plotted on a histogram, as in Fig. 6.8, then an association between repeat number and severity of disease becomes apparent: in general, the more severe the phenotype the larger the expanded allele. The minimally affected individuals have the smallest expanded alleles, of up to 0.3 kb (50–100 CTG repeats). Some of these individuals may have cataract as the only symptom, while others may appear entirely normal clinically.

The only significant difference found between the sexes is in this minimally affected class, where there is an excess of males. This may reflect a greater tendency to initial instability in male meiosis, rather than to an absolute excess of minimally affected males in the population. This sex difference may be of relevance when we discuss the mechanism by which it is possible for the CTG repeat number to increase (or expand) from an allele within the size range seen in the control population to that seen associated with DM. The other group of patients showing a marked relationship to size of repeat is the severely affected congenital group with expanded alleles of 2 to >6 kb (650–2000 CTG repeats) which, while it overlaps with the adult-onset group, is absolutely distinct from the minimal group.

There is also a strong correlation between apparent age at onset and the logarithm of repeat size, as shown in Fig. 6.9, with clinical features of the disease being detected at an earlier age in patients with a larger CTG repeat number. Thus, it has been clearly demonstrated that the size of the expansion observed in the genomic DNA of a patient affected by DM, is related to the DM phenotype. However, the correlation is not exact and the CTG repeat number cannot be used to predict accurately the clinical status of a patient.

INTERGENERATIONAL DIFFERENCES IN PHENOTYPE AND REPEAT SIZE: THE DM MUTATION IS DYNAMIC

If we look at parent–child pairs, the age at onset is generally earlier in the child than in the parent, while the repeat size is generally larger in the child than in the parent. This demonstrates that not only is the inverse relationship between size of repeat and age at onset maintained when the mutation is transmitted from parent to child, but also that the mutation is dynamic, with a tendency for the repeat number to increase when it is transmitted. A multicentre collaborative study (Ashizawa et al., 1994) has estimated that there is a 93–94% chance of an expanded allele further increasing in repeat number when it is transmitted. This provides a molecular explanation for the observed phenomenon of anticipation, or the earlier age at onset seen in affected children compared to their affected parent, as previously described.

On average, there is a greater increase in size of repeat when it is transmitted through females than when it is transmitted through males, although this can in part be explained by mothers having, on average, a larger

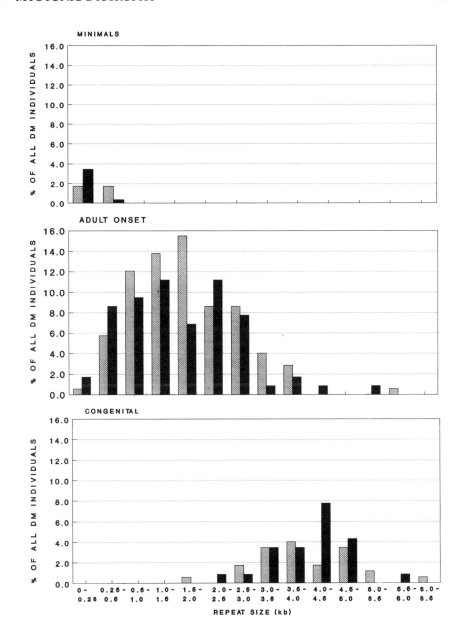

Fig. 6.8. CTG repeat sizes (in kb) for DM patients classified as minimal, adult onset or congenital according to the criteria described above under 'The myotonic dystrophy phenotype'. The results are expressed as the percentage of the total number of female (hatched box) or male (filled box) patients with repeat sizes as indicated on the x axis. A total of 301 patients were characterized

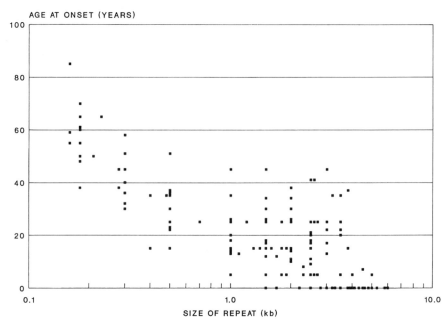

AGE AT ONSET (YEARS)

SIZE OF REPEAT (kb)

Fig. 6.9. Age of onset for 218 DM patients plotted against CTG repeat length (logarithmic scale)

repeat size than that seen in fathers. When expressed as a proportion of the size of repeat seen in the parent, the difference is no longer apparent. Fathers with a repeat size of greater than 2.0 kb rarely transmit a larger repeat to their children, whereas mothers with a repeat size of 2.0 kb or greater are able to transmit greatly expanded alleles. When the sperm of these males was analysed as a total pooled sample, no sperm with alleles larger than that detected in the blood were found; in fact the repeats detected were often smaller or no expanded repeats were detected at all. This suggests that the sperm in such an individual contained only an unexpanded chromosome 19. Alleles of > 1000 CTGs have not been detected in sperm, suggesting that large repeat alleles may greatly reduce the fitness of sperm carrying them, or may be detrimental to the progenitor cells such that only sperm bearing non-expanded repeat alleles will be produced. The larger repeats may also affect the fertility of the male in a *trans*-acting manner. Alternatively analysis of a total pooled sperm sample may not be the most sensitive method of identifying larger alleles. A feature of the disease in males is that those with larger CTG repeats are likely to have a low healthy sperm count, or be totally lacking in sperm.

   Although the general trend is for the repeat to increase in size from parent to child, it can be stably transmitted, particularly if it is small, or may even contract in size. These contractions preferentially occur in male trans-missions, with an estimated frequency of 6–7% (Ashizawa *et al.*, 1994). That such events have been detected preferentially in males may be a consequence of the observed detrimental effect that larger repeat sizes have on spermatogenesis and male fertility. Several cases have been reported where a reduction in repeat number has occurred on transmission from an affected parent with a large repeat to an offspring with a repeat size in the range seen in the control population. Furthermore, there are cases where intergenerational repeat length regression does not correlate with decreased severity of disease, again making it difficult to predict accurately the DM phenotype based on repeat size. The data showing these contractions have originated in several different laboratories. Differences in the way the data were generated and scored may have overestimated the actual number of contractions. If a more sensitive test was carried out, for example PCR on one or a few cells (known as small pool PCR or SP-PCR; Jeffreys et al., 1994), an overlap might be detected between the sizes seen in fathers and their children which might be missed in a less sensitive test.

The repeat size is usually measured in DNA extracted from blood. Blood DNA has been shown to have a lower level of mosaicism than DNA extracted from sperm, but SP-PCR has shown that despite differences in the extent of mosaicism, the lower level of repeats is the same in both sperm and blood. This would support the view of expansion from an initial common repeat number and suggest that contractions are rare and not usually seen in somatic cells, otherwise we would not expect to see an identical lower limit in two different cell populations but a statistically normal distribution of repeat sizes around a common mean within each tissue type. Not to be forgotten is the fact that we are looking at dynamic mutation which shows somatic mosaicism. Somatic expansion will have occurred in both the father and child since the time of conception. If the father transmitted a sperm with a repeat number at the very lower limit of that present in his sperm cells, and if somatic expansion has continued in the father since the conception of the child, thus taking his repeat sizes to a range larger than that of his child, this could explain why the child has repeat sizes smaller than and non-overlapping with those of the father. Although this explanation does require many assumptions, it illustrates that dynamic mutations are not easy to study. However, the fact still remains that these exceptional results have been seen at a much higher frequency in male than in female transmission.

The incomplete penetrance that was observed in many DM families prior to identification of the mutation can now be explained in terms of either transmission of an unaltered, small number of CTGs or by a contraction upon transmission.

THE MOLECULAR BASIS OF CONGENITAL MYOTONIC DYSTROPHY

The largest repeat sizes are seen in individuals who are congenitally affected, but with a considerable overlap between the range of repeats seen in this group and that seen in the adult-onset group. How can we explain the different phenotypes resulting from expanded alleles of a similar size? If the repeat size of these offspring is plotted against that of their affected parent as shown in Fig. 6.10, subdivided into paternal transmission (all offspring), maternal transmission (minimal or classical offspring) and maternal transmission (congenital offspring), an interesting observation is made. The maternal transmissions giving rise to congenital offspring form a separate group in terms of their distribution on the graph, from those giving rise to non-congenital offspring. This suggests that some interaction between the genotypes (and/or phenotypes) of the mother and infant is responsible for the more severe form of the disease, although it does not describe a precise relationship between the two.

In addition, when this type of analysis is performed, the DNA used is routinely extracted from white blood cells. It has been shown that the size of expansion seen in DNA extracted from skeletal muscle cells of a particular individual is often much larger than that seen in their white blood cells; thus if we were to look at the repeat sizes seen in the skeletal muscle of congenital DM patients, they might then all fall into a distinct group in terms of the size

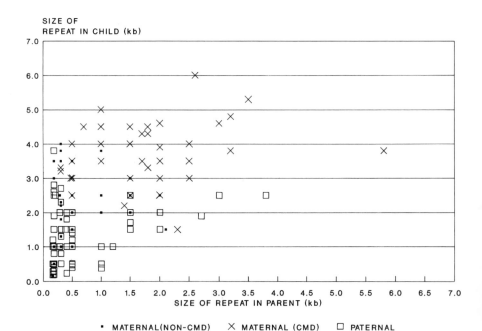

Fig. 6.10. Repeat size in parent–child pair, separated into maternally transmitted congenital onset, maternally transmitted non-congenital onset and paternally transmitted DM. In these three categories, the number of individuals was 46, 33 and 83 respectively

of repeat. It is not often that muscle biopsies are taken from patients (an unpleasant though not particularly painful procedure) because they are not essential for diagnosis in these more severe cases. Thus somatic mosaicism may be making the picture appear less clear than it really is.

Only two cases of an affected father and a normal mother giving birth to a congenital DM child have ever been reported. In the first case two affected offspring were born, the repeat size having increased from 0.6 kb (200 CTG repeats) in the father to 2.7 kb (900 repeats) and 4 kb (~1350 repeats) in his two daughters, who were diagnosed as having early-onset and congenital DM respectively. The second report was of a male with an expanded allele of 0.3 kb (100 repeats) transmitting an allele of 4.5 kb (1500 repeats) to a congenitally affected son. In both cases the father was mildly symptomatic. This could be taken as evidence that repeat size alone is sufficient to determine the phenotype. Perhaps we do not generally see congenitally affected infants born to affected fathers because of the effect that repeat sizes of >2.0 kb appears to have on male fertility.

DOES THE SAME MECHANISM OPERATE FOR EXPANSIONS AND CONTRACTIONS? WHEN DOES THE CHANGE IN REPEAT SIZE OCCUR?

Experiments comparing blood and sperm DNA from the same individual, although showing mosaicism in both tissues, demonstrated that there is greater variation in sperm than in blood. This may simply reflect the fact that some tissues will undergo more somatic divisions than others and thus the expanded allele will have many more chances to further increase in size in these tissues. Evidence from twin studies and from the pattern of mosaicism observed in a range of tissues from single individuals suggests that the somatic expansion is established in early embryogenesis. Studies of other simple sequence repeat polymorphisms support the conclusion that allelic variation in an individual occurs during the first few cell divisions after fertilization (Gibbs *et al.*, 1993). Cases have been observed where the length of CTG repeat was much larger in the blood of the offspring than in the sperm of the father, suggesting that repeat expansion has occurred during development or growth of the offspring. SP-PCR should confirm whether or not there is overlap between the size of alleles seen in these parent and child pairs.

A general observation has been made that the repeat size can vary from cell to cell (detected as a smeared band on an autorad), and from tissue to tissue (referred to as somatic mosaicism) within an individual. Thus we can conclude that the change in repeat size can occur after replication of the chromosome in somatic cells as well as in the production of germ cells. Recombination of flanking markers is not seen when either expansion or contraction of repeat number occurs, confirming that what we observe happens during replication of the chromosomes, rather than because of something specific to meiosis, such as an error in the recombination process caused by a large number of CTG repeats.

EVIDENCE THAT ALL CHROMOSOMES WITH AN EXPANDED CTG REPEAT NUMBER ARE DESCENDED FROM A COMMON ANCESTOR: ONE ORIGINAL MUTATION?

Even before the DMPK gene was cloned and the DM phenotype was shown to be the result of an increase in the number of CTG repeats present in the 3'-UTR of the gene, there was good evidence that every person with DM was descended from one or a very few common ancestors. The first evidence for this came from the study of a polymorphism at the locus D19S63, which was shown to be very closely linked to the DM locus (0% recombination). The polymorphism consists of three alleles detected in DNA digested with the restriction enzyme PstI. When inheritance of the polymorphism was studied in DM families, thus enabling identification of which allele was co-inherited with the disease gene, the frequencies of the three alleles on DM chromosomes were found to be very different from that determined in a control population. Allele 3 was found to be co-inherited with DM at a much higher frequency than would be predicted if there was random association of the two loci (Harley et al., 1991). Strong association (sometimes referred to as linkage disequilibrium) between particular alleles at two polymorphic loci, in this case allele 3 of D19S63 and the DM mutation as opposed to the normal DMPK gene, indicates that such loci were associated as part of a haplotype which arose many generations ago and has subsequently been inherited as a unit because of their physical closeness.

Following this, a second polymorphism (locus D19S95) was identified which showed complete association with DM; that is, all DM chromosomes were inherited with the same allele. The polymorphism was detected because some chromosomes had an extra 1 kb of DNA, the alleles being referred to as either insertion (with the extra 1 kb) or deletion (lacking the extra 1 kb) alleles, the insertion allele always being inherited with DM. When the inheritance of this polymorphism and the number of CTG repeats in the DMPK gene was studied, an interesting pattern of association was discovered. Chromosomes with either five or >18 CTGs were found almost always to be co-inherited with the insertion allele at D19S95. Conversely, chromosomes with 11, 12 or 13 CTGs were almost always co-inherited with the deletion allele. It had already been demonstrated that the insertion allele was always inherited with DM chromosomes, i.e. those with a repeat size >50. This observation led to the hypothesis that the DM-causing chromosomes had descended from chromosomes with repeat sizes of >18, which in turn may have originated by an original mutation of a chromosome with five CTGs increasing to >18 CTGs in one step (Imbert et al., 1993; unpublished studies, this laboratory).

We know that large CTG repeat numbers cause instability leading to almost inevitable expansion at a rate dependent on the number of repeats. Smaller repeats within the normal size range may also be subject to expansion but at a much lower frequency than those seen in DM patients. If the $(CTG)_5$ allele was a predecessor of $(CTG)_{>18}$ caused by a rare initial expansion in one or a few alleles, these larger alleles would be more likely to undergo further expansion and eventually reach a size big enough to result in a DM phenotype. If this is true, then we have a simple answer to our question 'Why is DM not decreasing in frequency?' If CTG alleles with >18 repeats are slowly expanding over many generations until they reach a size large enough to be detected because they cause phenotypic changes, we have a 'reservoir' of chromosomes within the general population which, given enough time, will ultimately replenish the pool of DM chromosomes being lost because their phenotypic effect is so severe that they have a reproductive fitness of zero.

As an alternative to the above, the existence of a haplotype predisposing to new mutation (i.e., expansion) could also explain the observations. Alleles of >18 might share some linked sequence, for example the 1 kb insertion allele, that predisposes them to further expansion and thus acts as a reservoir for apparently new DM mutations. However, DM has been identified in one family which lacks the insertion allele, suggesting that it is the CTG repeat number alone that drives the expansion.

Inspection of large DM families shows a significant excess of non-manifesting or minimally affected males in the top generation. If new alleles are being generated, then this observation suggests that they are more likely to arise after male transmission. This may reflect the greater number of cell divisions which occur during spermatogenesis as opposed to oogenesis.

Meiotic drive, or segregation distortion, has been demonstrated such that, in an individual heterozygous for two CTG alleles within the normal range, the larger of the two is preferentially transmitted. Presumably larger alleles within the normal range confer some selective advantage. This may have contributed to the rapid spread of alleles with >18 repeats in various populations.

NEW MUTATIONS OR PHENOCOPIES?

Five families have been identified which have clinical features of DM but do not have abnormal expansion of the CTG repeat. As yet it has not been possible to identify the specific change in DNA sequence responsible for the disease in these families, but it may well be that they have mutations in a different gene. The argument for this being the case, rather than there being other mutations within the DMPK gene, is as follows. Anticipation has also been observed in these exceptional families. This phenomenon is the result of a highly specific type of molecular change in the DNA. Other diseases caused by expansion of a triplet repeat can also be shown to exhibit some degree of anticipation related to the molecular expansion of the repeat. It would be surprising if some other type of molecular event could result in the same clinical observation of anticipation. However, as the 3'-UTR is thought to play some role in regulation of transcription or accurate processing of the mRNA, it is possible that a different molecular event within this portion of the gene could have a similar effect.

PRE-NATAL DIAGNOSIS AND PRE-SYMPTOMATIC TESTING

The different behaviour of the sequence in male and female meiosis, in particular the higher rate of contractions in male transmissions, clearly has implications for pre-natal diagnosis. Pre-natal diagnosis from chorion biopsy has so far proved reliable, especially in families where the congenital disease has occurred in a previously affected child. The outcome in male transmissions may be more variable because of the higher rate of contraction of repeat number. Only two cases of paternally transmitted congenital DM have been identified, thus giving a more favourable prognosis for male transmission.

Carrier detection and exclusion have also been fairly straightforward. Knowledge of the precise mutation causing the disease means that a molecular diagnostic test can be performed on an individual without the need for DNA samples from other family members. Previously identification of the inheritance of the chromosome carrying the DM mutation relied on use of linked markers, and a study of the pattern of their co-inheritance within a family. Use of linked markers to predict gene carriers is always accompanied by a low level of misidentification because recombination between the linked marker and the disease gene will alter their pattern of co-inheritance.

Ethical problems will always arise once it is possible positively to identify disease gene carriers, particularly in the case of a disease for which there is, as yet, no means of prevention or treatment. Who should be tested? For example, it has been shown that people with early-onset cataract have an increased risk over that of the general population of having a CTG repeat size at the lower end of the range found in minimal or asymptomatic DM carriers. Therefore, do early-onset cataract patients represent a group of individuals who should be genetically tested?

How confidential is genetic information? Who should have access to such information—the individual for whom it was obtained? Other family members? Will such testing result in insurance companies demanding testing of high-risk individuals before a life or health insurance policy is considered? These types of issues are the subject of much discussion within the medical community (Ball et al., 1994). It is likely that the ethical problems surrounding genetic testing will diminish once therapeutic methods have been developed for DM. This is still some way in the future as a full understanding of the fundamental biology underlying DM phenotype will be necessary before therapeutic strategies can be developed. The next section considers the current state of our understanding of the molecular basis of DM.

THE MOLECULAR BASIS OF MYOTONIC DYSTROPHY

The identification of an expanded trinucleotide DNA sequence as the mutational event in DM (Brook *et al.*, 1992b) represents a crucial step towards determining the molecular basis underlying the pathophysiology of this disorder. How the expanded repeat exerts its effect in this disease is the key question currently being addressed by DM researchers. Can anything be learned from other disorders caused by expanded triplet repeats?

OTHER DISEASES ARE CAUSED BY THE EXPANSION OF
TRINUCLEOTIDE DNA SEQUENCES

To date nine different disorders have been identified which are associated with the expansion of trinucleotide DNA sequences. The other eight are: spinal and bulbar muscular atrophy (SBMA); fragile X mental retardation syndrome (FRAXA); spinal and cerebellar ataxia type 1 (SCA1); Huntington's disease (HD); FRAXE (associated with mild mental retardation); Machado–Joseph disease (MJD); dentatorubral and pallidoluysian atrophy (DRPLA); and the allelic Haw River syndrome (HRS). In addition two other expanded repeats, FRAXF and FRA16A, have been identified but are not associated with any disorder. Although similarities exist between some of the disorders and DM, in terms of the type of triplet involved and the extent of repeat

Table 6.3. Expanding triplet repeats in human disease

	Triplet	Repeat size range		Location
		Normal	Disease	
Kennedy's disease (SBMA)	CAG	13–30	40–62	Androgen receptor (coding)
DM	CTG	5–34	50 > 2000	3'-UTR DMPK
HD	CAG	6–34	36–121	ORF
SCA1	CAG	25–36	43–81	Prob. ORF
DRPLA/HRS	CAG	7–23	49–75	ORF
MJD	CAG	13–36	68–79	ORF
FRAXA	CGG	7–52	50–2000	5'-UTR FMR1
FRAXE	CGG/GCC	6–25	50 > 2000	?
FRAXF	CGG/GCC	6–29	300–5000	?
FRA16A	CGG/GCC	16–49	1000–2000	?

expansion, none of these cases is exactly the same as DM. For SBMA, SCA1, HD, DRPLA/HRS and MJD the expanded sequence is CAG/CTG as in DM. However, the extent to which the repeats are expanded in each of these disorders is considerably less than in DM. Table 6.3 shows the range of repeat lengths at each of these loci in the normal population and in affected individuals. The triplet repeats in SBMA, SCA1, HD, DRPLA/HRS and MJD are located within the open reading frames (ORFs) of transcripts and are therefore coding. It is possible that in each of these disorders the length of repeat expansion may be constrained by the effect additional amino acids would have on functional proteins, which in turn may affect offspring viability. Hence the extent to which repeat expansion can be tolerated and appear in viable offspring is limited. No such constraints apply to a situation seen in DM where the triplet repeat is located in the non-coding 3'-UTR.

For FRAXA, FRAXE, FRAXF and FRA16A the expanded repeat is a different trinucleotide from that in DM. In these cases the triplet is CGG/GCC. However, the extent of repeat expansion is comparable to that seen in DM, i.e. expansions of several kilobases are not uncommon. In FRAXA, similar to the situation in DM, the expanded repeat is not translated. In this case the trinucleotide repeat is located in the 5'-UTR of the FMR1 gene. For FRAXE, which is associated with mild mental retardation, the location of the expanded repeat with respect to coding sequences has yet to be identified. FRAXF and FRA16A are not known to be expressed nor are they associated with a phenotype.

In FRAXA it has been shown that methylation of the expanded repeat results in the switching off of the FMR1 gene and an absence of transcripts from this locus. No such mechanism operates in DM, however, as studies have shown that imprinting is not involved in the expression of DMPK, and in any case the CTG repeat does not contain the methylatable sequence CpG.

QUANTIFICATION OF DMPK EXPRESSION

So what is happening in DM? Four studies have reported on RNA quantification of DMPK in DM tissues, with contradictory results. One study indicated that DMPK mRNA levels are increased due to elevated levels of mutant transcripts. This finding is at odds with three other studies which showed that the level of DMPK RNA is reduced in DM patients due to a lack of expression from the mutant allele. There are significant differences in the tissues and cells selected and in the procedures employed for RNA quantification in each of these studies, and clarification of these findings is awaited. It should be noted, however, that the only study published to date on DMPK protein quantification indicates that level of DMPK protein is reduced in DM patients (Fu et al., 1993).

An alteration in the level of DMPK may result from an abnormality of processing or synthesis of the transcript from the abnormal allele. An example of such a mechanism is the double-sex mutant of *Drosophila* where mRNA processing is defective due to mutation in a repeat sequence which normally binds protein. Similarly in *Caenorhabditis elegans* mutations in the 3'-UTR of the *fim-3* and *lin-14* genes producing unregulated activity is thought to result from deletion or inactivation of a binding site for a negative control element. Such an effect may result from CTG amplification in DM.

The importance of 3' untranslated sequences has been demonstrated recently in two different studies. In one study the 3'-UTR was shown to be important in directing different actin mRNAs to the appropriate cellular compartment. Other studies have shown that the 3'-UTRs of troponin 1, tropomysin and α-cardiac actin are *trans*-activating regulators in a feedback loop promoting differentiation and inhibiting division. Such a mechanism involving the 3'-UTR of DMPK might indicate that effects on protein kinase levels are less important to the DM phenotype than the interaction of the 3'-UTR with other cellular factors. There is currently little direct evidence to support this proposition, though a recent report using gel shift assays indicates that DNA binding proteins interact with $(CTG)_n$ repeats.

BIOCHEMICAL ABNORMALITIES IN MYOTONIC DYSTROPHY

How do these data correlate with what is already known about the biochemical basis of DM? Several biochemical abnormalities have been documented in DM, though their relevance is unclear. These include: an abnormal stoichiometry of sodium and potassium transport, with a reduction of sodium extrusion in relation to potassium uptake; a reduction of 30–60% of the activity of adenyl cyclase in sarcolemmal membranes of DM patients; and abnormalities in calcium transport resulting from a change in membrane permeability. Perhaps the most relevant biochemical study was made 20 years ago when the role of protein kinases in DM was suggested. Experiments indicated that there was a reduction in the phosphorylation of red blood cell membranes in DM patients compared with controls (Roses and Appel, 1973). This study was extended to sarcolemmal membranes and showed that in DM patients the phosphorylation of proteins at a molecular weight of around 30 000 and 50 000 was roughly half the normal level. The significance of these findings is unclear, though studies on the protein kinase, particularly those to identify the substrate or substrates on which it acts, should prove illuminating. Clearly this area of research is likely to provide valuable insights into DM over the next few years.

COULD THE EXPANDED REPEAT AFFECT FLANKING GENES?

In view of the contradictory evidence concerning the level of DMPK RNA in DM patients it is worth considering whether the expanded triplet affects the expression of other genes adjacent to DMPK, or whether the ratio and distribution of DMPK alternative splice forms may be misrepresented in the mRNAs of different cells. To address the first of these issues it is necessary to identify other transcripts in the vicinity of the expanded CTG. During the search for the DM locus several transcripts and putative transcripts were identified. A transcript map of the interval flanking the triplet repeat is shown in Fig. 6.11. Five cDNAs have been identified in an interval of 50 kb and all are transcribed in the same orientation. Of these three transcripts 20-D7 is the least well characterized. As 20-D7 is transcribed in the same direction as DMPK and 59, it is possible that the expanded repeat exerts an effect on the 20-D7 promoter region. Quantification experiments on RNA levels transcribed from 20-D7 are currently in progress to address this issue.

The two transcripts most closely associated with the DM triplet are DMPK and 59. As stated earlier, the triplet repeat is located in the 3'-UTR of DMPK. Although it was known from mapping data that 59 is located close to DMPK, it was not until the genomic DNA surrounding each of these genes was sequenced that their immediate proximity became apparent. The two genes are in the same transcriptional orientation and the 5' end of the longest DMPK cDNA is located 502 bp from the 3' end of the longest 59 cDNAs. This stretch of DNA contains putative binding sites for several different transcription factors. Clearly, characterization of this interval as the likeliest promoter region for DMPK should provide valuable information for our understanding of DMPK regulation.

Human and mouse clones for DMPK and 59 have been characterized. Northern blot data on a variety of rodent tissues indicate that the patterns of expression of DMPK and 59 are complementary; that is, in general tissues where 59 is most strongly expressed have lower levels of DMPK and vice versa (Jansen et al., 1992). This has led to the suggestion that 59 may be involved in DM, particularly as so many different tissues are affected in this

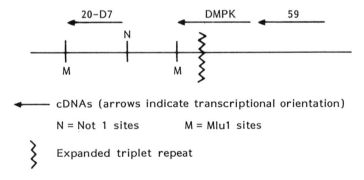

Fig. 6.11. Location and transcriptional orientation of three genes, 20-D7, DMPK and 59, within a 50 kb interval surrounding the expanded repeat

disorder, including those in which DMPK is poorly expressed. In brain, for example, DMPK is expressed at very low levels whereas 59 is fairly abundant. Quantification of 59 RNA levels should provide data to confirm or refute the involvement of 59 in DM. In the absence of good evidence to the contrary, a direct effect on DMPK expression by triplet repeat expansion, possibly associated with the perturbation of other cellular factors, remains the likeliest molecular mechanism causing the DM phenotype.

DMPK AND ITS GENOMIC ORGANIZATION

Detailed characterization of DMPK reveals that the full-length mRNA is about 3.3 kb in size. The translated region consists of 618 amino acids and the putative protein kinase domain is encoded by residues 69–338. A domain predicted to have a high α-helical content is located between amino acids 424 and 534 and at the C-terminal end of the sequence is a region of high hydrophobicity, possibly indicating a membrane-bound domain. Fig. 6.12 shows the predicted structure of DMPK. The genomic organization of DMPK has been worked out and it consists of 15 exons spanning 12 kb of genomic DNA. Several alternative splice forms have been identified, principally involving exons 13 and 14, though forms that are variable at the 5' end have also been documented (Fig. 6.13). There is also another splice form containing an additional 15 bp which results from alternative splice site selection at the 3' end of exon 8. The functional significance of these alternative forms remains to be characterized.

Several groups are raising and characterizing antibodies to the DMPK protein. One study has reported antibodies detecting a 53 kDa protein which is present at low levels in skeletal and cardiac muscle extracts of DM patients and controls. Furthermore, examination of muscle sections reveals that DMPK localizes prominently to neuromuscular and myotendinous junctions of human and rodent skeletal muscles. Interestingly, this study also presented preliminary data showing that the protein was present at the neuromuscular junction in muscle tissues from adult and congenital DM cases. These studies are still at an early stage but clearly detailed knowledge of the DM protein kinase and its substrate are likely to be a prerequisite for understanding the molecular basis of DM.

IN CONCLUSION

The development of molecular cloning techniques has transformed human genetics from a subject of clinical observation and pedigree collection to a stage where the genes responsible for many of the common inherited disorders have been identified and cloned. Over the past 10 years research into myotonic dystrophy has progressed from the chromosomal assignment

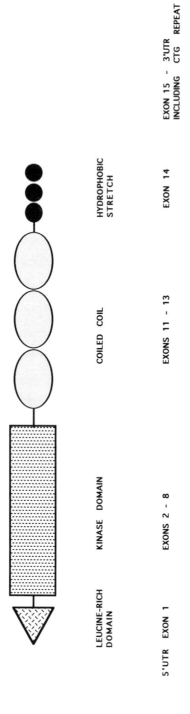

Fig. 6.12. Representation of the likely structure of the DMPK protein and how each of the domains relate to the various exons of the gene

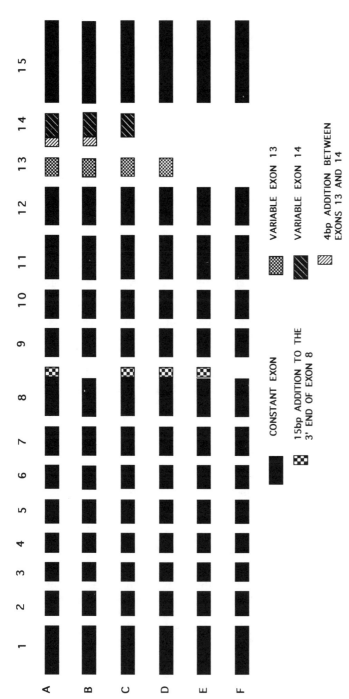

Fig. 6.13. Six alternative splice forms of the myotonic dystrophy protein kinase gene (AnF). The number of each exon is listed at the top of the figure

of the locus to the cloning of the DM gene and identification of the mutational mechanism. The challenges of the next 10 years are clear. A firm understanding of the fundamental biology underlying this disorder will be necessary in order to address the ultimate challenge for DM researchers, i.e. the development of strategies for therapeutic intervention.

ACKNOWLEDGEMENTS

We wish to thank Mrs R.J. Goodwin for assistance with the manuscript. Work in our laboratories is funded by the Muscular Dystrophy Association (USA) and the Muscular Dystrophy Group of Great Britain.

REFERENCES

Ashizawa T, Anvret M, Baiget M et al. (1994) Characteristics of intergenerational contractions of the CTG repeat in myotonic dystrophy. Am. J. Hum. Genet., 54, 414–423.

Ball D, Tyler A and Harper P (1994) Predictive testing of adults and children. In Clarke A (ed.), Genetic Counselling: Practice and Principles, 1st edn. London: Routledge.

Brook JD, Zemelman BV, Hadingham K et al. (1992a) Radiation-reduced hybrids for the myotonic dystrophy locus. Genomics, 13, 243–250.

Brook JD, McCurrach ME, Harley HG et al. (1992b) Molecular basis of myotonic dystrophy: expansion of a trinucleotide (CTG) repeat at the 3' end of a transcript encoding a protein kinase family member. Cell, 68, 799–808.

Davies KE, Jackson J, Williamson R et al. (1983) Linkage analysis of myotonic dystrophy and sequences on chromosome 19 using a cloned complement 3 gene probe. J. Med. Genet., 20, 259–263.

Fu Y-H, Friedman D, Richards S et al. (1993) Decreased expression of myotonin-protein kinase messenger RNA and protein in adult form of myotonic dystrophy. Science, 260, 235–238.

Gibbs M, Collick A, Kelly RG and Jeffreys AJ (1993) A tetranucleotide repeat mouse minisatellite displaying substantial somatic instability during early preimplantation development. Genomics, 17, 121–128.

Harley HG, Brook JD, Floyd J et al. (1991) Detection of linkage disequilibrium between the myotonic dystrophy locus and a new polymorphic DNA marker. Am. J. Hum. Genet., 49, 68–75.

Harley HG, Brook JD, Rundle SA et al. (1992) Expansion of an unstable DNA region and phenotypic variation in myotonic dystrophy. Nature, 355, 545–546.

Harper PS (1989) Myotonic Dystrophy, 2nd edn. London: WB Saunders.

Imbert G, Kretz C, Johnson K and Mandell J-L (1993) Nature Genet., 4, 72–76.

Jansen J, Mahadevan M, Amemiya C et al. (1992) Characterisation of the myotonic dystrophy region predicts multiple protein isoform-encoding mRNAs. Nature Genet., 1, 261–266.

Jeffreys AJ, Tamaki K, MacLeod A et al. (1994) Complex gene conversion events in germline mutation at human minisatellites. Nature Genet., 6, 136–145.

Roses AD and Appel SH (1973) Protein kinase activity in erythrocyte ghosts of patients with myotonic muscular dystrophy. Proc. Natl Acad. Sci. USA, 70, 1855–1859.

Shaw DJ, Meredith AL, Sarfarazi M *et al.* (1985) The apolipoprotein C11 gene: sub-chromosomal localisation and linkage to the myotonic dystrophy locus. *Hum. Genet.*, **70**, 271–273.

Shutler G, Korneluk RG, Tsilfidis C *et al.* (1992) Physical mapping and cloning of the proximal segment of the myotonic dystrophy gene region. *Genomics*, **13**, 518–525.

FURTHER READING

Brook JD (1994) Positional cloning. *Scientific American* (Science and Medicine), **1**, 48–57.

Clarke A (1994) *Genetic Counselling: Practice and Principles*, 1st edn. London: Routledge.

Wieringa B (1994) Commentary. Myotonic dystrophy reviewed: back to the future? *Hum. Mol. Genet.*, **3**, 1–7.

7 Genetic and Molecular Analysis of the Human Y Chromosome

JEAN WEISSENBACH

Our knowledge on the genetic control of sex determination has progressed in fits and starts since the beginning of the century. It was first noticed, in a number of species, that in individuals of one sex (but not the other) one or several chromosomes could not be associated by pairs in an otherwise diploid karyotype. The term 'sex chromosomes' was then used to designate chromosomes distributed in a sex-specific manner. It was inferred that sex determination resulted from the distribution of sex chromosomes. However, in many species sex determination is triggered by a non-genetic factor and in such species karyotyping does not reveal sex chromosomes.

In mammals, it has long been thought that sex was determined by the X: autosome ratio, as in the fruit fly. But since the late 1950s we have known that sex determination in mammals is controlled by the Y chromosome. In the presence of a Y chromosome the primordial undifferentiated gonads develop as testes whereas in its absence they develop as ovaries. On this basis it has been concluded that the Y chromosome carries at least one gene which controls primary sex differentiation. It was even suggested that the Y chromosome would not harbour any further function. The term 'testis-determining factor' (TDF) is presently used to designate this gene. Cases of sex reversals in which the observed gonadal sex is in contradiction to the karyotype began to be observed as cytogenetics became more accurate. As there are two sexes, there are also two main classes of sex reversals: females with a 46,XY karyotype (XY females) and males with a 46,XX karyotype (XX males). It was then proposed that an abnormal interchange between the distal parts (more remote from the centromere) of arms of the human X and Y chromosomes could mobilize the sex-determining region of the Y chromosome, transfer it to the X and result in both types of sex reversals. Cases of sex reversal were also observed in other mammals and a sex reversal mutation (Sxr) transmitted through male carriers was identified in mouse. In humans, with a few exceptions, sex reversal is accompanied by sterility precluding genetic linkage analysis. However linkage analysis was not necessary to show that TDF is located on the Y chromosome, and in any case most of the latter

Molecular Genetics of Human Inherited Disease. Edited by D.J. Shaw
Published 1995 by John Wiley & Sons Ltd

cannot be mapped by genetic linkage analysis. With the improvement of techniques in cytogenetics it also became possible to define more accurately chromosomal rearrangements and to try to correlate them with cases of partial or complete sex reversal. Several cytogenetic analyses suggested the first tentative locations of a sex-determining region on the Y chromosome. These observations usually but not always mapped the sex-determining region to the short arm (Yp) and in general to its proximal part (closer to the centromere). DNA analysis of the Y chromosome in sex reversals led to the exact localization and subsequent identification of TDF in man and its homologue, Tdy, in mouse. This identification took 8 years since the isolation of the first DNA single-copy probes specific for the Y chromosome. Though this review is focused on the human Y chromosome, it will refer frequently to the mouse. Molecular and genetic studies of sex determination in each species have been strengthening one another and this interplay accelerated the increase of knowledge during the past decade. This huge molecular mapping effort has also resulted in the identification of other Y-specific genes, the functions of which remain largely hypothetical or unknown, but for the study of which again the mouse model will be crucial.

THE Sxr MOUSE MODEL

The oldest and most extensively studied sex reversal condition in the mouse is Sxr or 'sex-reversed'. Sxr causes chromosomal XX and XO females to develop as phenotypic male mice. It is transmitted by XY Sxr carrier males as an apparent autosomal dominant trait: a cross between normal XX females and XY Sxr male carriers results in F1 progeny of 1/4 of normal XX females, 1/4 of XX Sxr males, 1/4 of XY Sxr carrier males and 1/4 of normal XY males, i.e. 3/4 of phenotypic males (Fig. 7.1a).

The first series of observations was made in mouse with a satellite DNA probe (Bkm) isolated from the W sex chromosome of snakes which detected a Y-specific DNA fragment in some DNA digests in mouse. This Y-specific band was observed on Southern blots from XX Sxr mice. In situ hybridization of Bkm showed an abundance of this repeated sequence on the tip of one large chromosome in XX Sxr mice, presumably the paternal X, and in two regions on a small chromosome of XY Sxr male carriers, presumably the Y. Normal males showed only one such region on this small chromosome and normal females none. The presence of a second Bkm region on the tip of the Y of Sxr carriers was supposed to result from a duplication of the Bkm-containing chromatin of the normal Y chromosome (Fig. 7.1b). This implied that the sex-determining gene Tdy mapped within the Bkm region. Because of the sex-specific location of this repeated sequence in other vertebrates, it was even suggested that Bkm itself was involved in sex determination. The segregation of Sxr in the offspring was ascribed to a meiotic interchange

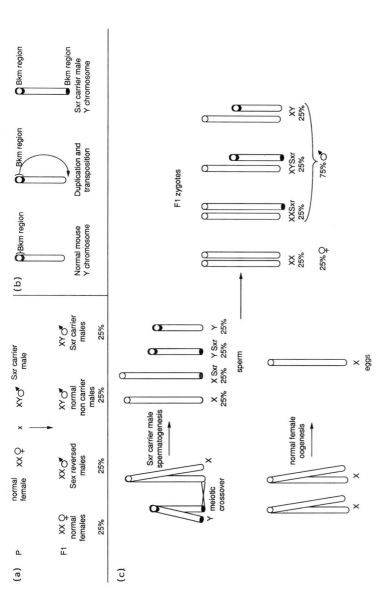

Fig. 7.1. Segregation of the Sxr mutation in an F1 generation of a cross between an XY 'sex-reversed' carrier male and a normal female. (b) The Sxr mutation has possibly occurred through a rearrangement of the mouse Y chromosome involving a duplication and a transposition of the pericentromeric Bkm region. Note that the mouse Y chromosome bears a very short arm whereas other mouse chromosomes are completely acrocentric. (c) Segregation of Sxr in gametes produced in cross of panel (a). An obligatory crossing over involving one chromatid of each sex chromosomes in male meiosis produces four different types of gametes with an equal frequency

through crossing-over between distal parts of a Y chromatid (mobilizing the duplicated Bkm segment) and an X chromatid (Fig. 7.1c). The cytogenetic proof of this model was provided soon after. These first molecular observations were an encouragement for similar studies in man, in whom sporadic cases of sex reversals are better documented.

PRIMARY MAPPING OF TDF

The first studies based on DNA analysis in man used repeated DNA probes from the heterochromatin of the distal part of the long arm, Yq12 (Fig. 7.2), and confirmed that these sequences were not involved in sex determination. Single-copy DNA probes from the Y chromosome became available in the early 1980s and were used to analyse the more abundant XX males. Human XX maleness is essentially sporadic, it is the most frequent sex reversal

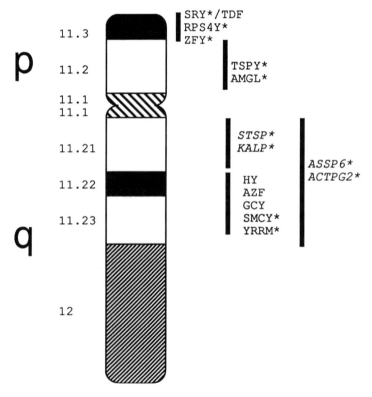

Fig. 7.2. Diagram of the human Y chromosome with mapped Y-specific genes and pseudogenes. Cloned genes and pseudogenes are marked with an asterisk and pseudogenes italicized

syndrome in man and occurs in about 1 in 20 000 males. XX males develop testes in the absence of a cytogenetically detectable Y chromosome, have an apparently normal male phenotype, but are sterile with small azoospermic testes. Presence of Y chromosomal material on the tip of the X chromosome short arm, a possible product of an abnormal X–Y interchange, was first convincingly suggested by a cytogenetic study on prometaphase bandings. Occurrence of Y-specific DNA in XX males was then demonstrated using Y-specific DNA probes (Guellaën *et al.*, 1984). Such XX males are designated hereafter as Y(+) XX males. This finding paved the way to deletion mapping of TDF on the basis of its presence on the Y chromosomal segment found in XX males. The smallest Y chromosome segment common to these patients defined the interval containing TDF (Fig. 7.3). In parallel with these mapping studies, it was shown that XX males inherited one X chromosome from their father and Y-specific DNA was detected by *in situ* hybridization at the short

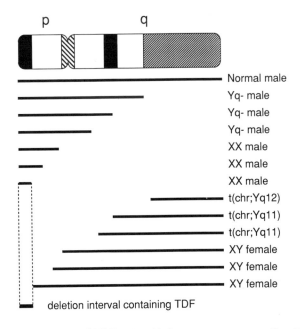

Fig. 7.3. Deletion mapping of TDF using Y chromosome anomalies. The horizontal bars represent the chromosomal segment present in a genome as defined by DNA analysis using Y-specific probes. Yq– are deletions of the long arm of the Y chromosome; t(chr;Yq...) are translocations, onto another chromosome, of Y chromosome segments defined by their breakpoint. XX males are Y(+) XX males with a Yp chromosome segment and XY females are cases with a Yp deletion. Overlap of the segment deleted in XY females and present in XX males defines the minimal interval containing at least a part of TDF

arm telomere of one of the X chromosomes of XX male patients. A deletion in the Y chromosome short arm could also be observed in some XY females. Molecular analyses showed that this deletion overlapped the Y chromosomal segment present in XX males (Fig. 7.3). More recently some other sex chromosome aneuploidies with sex reversal like 45,X males and 47,XXX males could be explained on the basis of a translocation of a Y chromosome segment onto an autosome in most 45,X males, or onto an X chromosome in one 45,X male and in 47,XXX males.

Similarly other functions encoded by the human Y chromosome have been mapped using panels of deletions associated with phenotypes. Recently, the deletion mapping procedure was further extended to the entire chromosome using sequence tagged sites (STSs) that can be readily characterized by a PCR amplification. These STSs could be ordered in a total of 43 intervals and served to order a set of contiguous overlapping yeast artificial chromosomes (YACs) covering the euchromatic part of the human Y chromosome (Foote *et al.*, 1992).

THE PSEUDOAUTOSOMAL REGION AND X–Y INTERCHANGE IN XX MALES

Isolation of DNA probes from the subtelomeric region of Yp showed that the terminal part of the short arms of the human X and Y chromosomes share a region of completely homologous DNA sequences (Fig. 7.4). Moreover, loci from this region can undergo recombination between both sex chromosomes during male meiosis (Fig. 7.5a). Because genetic linkage of this region to sex is either absent or partial, it was called the pseudoautosomal region. The mammalian Y chromosome can thus be divided into two functional segments. The pseudoautosomal segment, located in a terminal part, undergoes an obligatory crossing-over with a terminal part of the X chromosome during male meiosis. In eutherians, this crossing-over is necessary to ensure proper segregation of the X and Y chromosomes in separate sperms. The second segment is specific to males, bears functions required for male development and fertility and contains the centromere. The size of the human pseudoautosomal segment was estimated at 2600 kb, and it represents the

Fig. 7.4. (Opposite) Regions of DNA sequence homologies shared by the human X and Y chromosomes. Arrows indicate regions sharing sequence homologies which are represented by vertical bars along the chromosome diagrams. The pseudoautosomal region contains several genes which can undergo recombination between the X and Y during male meiosis. AMG, RPS4Y/X, ZFY/X and SMCY/X are chromosome-specific genes present on both sex chromosomes. STS and KAL are X-specific genes homologous to Y-specific pseudogenes. Arrows 1, 2 and pseudoautosomal region 2 represent highly homologous regions without known genes and translocated during recent human evolution. For a review on X–Y homologies see Affara and Ferguson-Smith in Wachtel (1994)

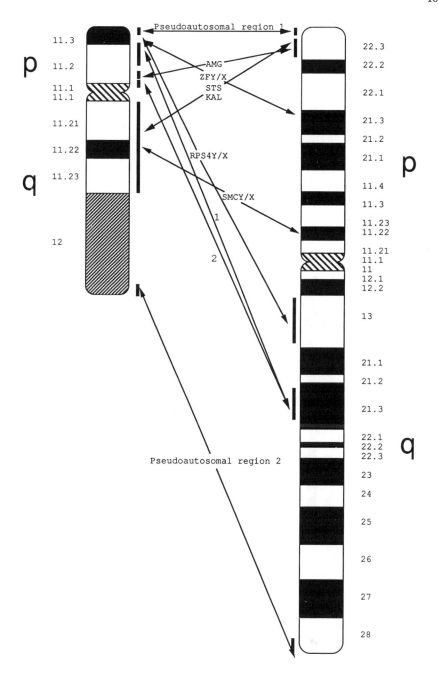

Fig. 7.4.

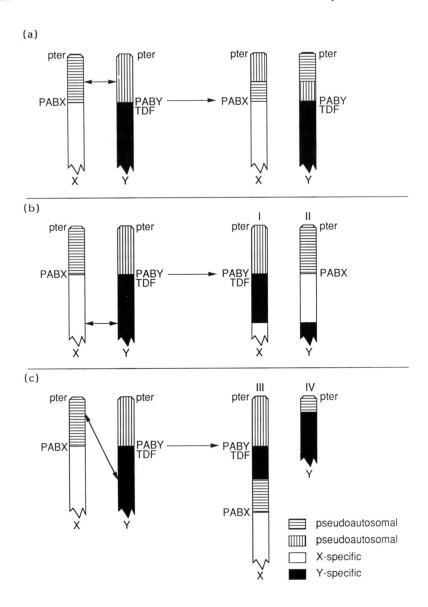

Fig. 7.5. Meiotic X–Y interchanges. Only the terminal parts of the short arms have been represented. Double arrows indicate the exchange points on each chromosome. Loci distal to exchange points cross over from one chromosome to another. (a) Normal X–Y crossing over. (b, c) Abnormal X–Y interchanges. X chromosomes I and III are found in XX males. Y chromosomes II and IV are found in XY females. IV can also result from a normal X–Y crossing over followed by an interstitial deletion of the Y. PABX: pseudoautosomal boundary of the X chromosome; PABY: pseudoautosomal boundary of the Y chromosome; pter: short arm telomere

terminal part of chromosome band Yp11.3 (Fig. 7.4). A number of genes have recently been identified in the pseudoautosomal region of the human Y chromosome and were shown to escape X inactivation.

Deletion mapping had located TDF to the distal interval of the short arm (Fig. 7.3). This interval could then be subdivided into two parts by a boundary (termed the pseudoautosomal boundary, PAB) separating the pseudo-autosomal region from a sex-specific region containing TDF. Existence of the pseudoautosomal region and its genetic properties also lend strong support to the interchange mechanism in the aetiology of sex reversal, which could originate from an abnormal X–Y interchange. However, although very suggestive, the *in situ* hybridization results mentioned above could still not indicate whether the paternal X chromosome of Y(+) XX males actually resulted from an interchange involving the terminal part of both paternal sex chromosomes or if this interchange was of internal material, leaving the telomeres on their original chromosomes. Family analyses of pseudoautosomal loci in Y(+) XX males indicated presence of the entire pseudoautosomal region from the paternal Y chromosome and absence of the terminal part of the short arm of the father's X (Petit *et al.*, 1987). These results definitely confirmed the terminal X–Y interchange model in which a translo-cation (or unequal crossover) occurs proximal to TDF and causes the transfer of the distal part of Yp onto the tip of Xp (Fig. 7.5b, c). The breakpoint on the X chromosome occurs either proximal or distal to the pseudoautosomal boundary PABX (Fig. 7.5b, c). In many instances deletion of X-specific loci from Xp22.3 can be observed.

Another pseudoautosomal region has been identified at the tip of the long arms of the human X and Y chromosomes. However, recombination between the long arm pseudoautosomal regions is not obligatory and occurs in only a fraction of meioses. In great apes this region is only found on the X chromosome. It is supposed that the region has been transferred onto the terminal part of Yq during recent human evolution. The size of this second pseudoautosomal region has recently been estimated at 320 kb.

X–Y INTERCHANGE IN XY FEMALES

As already mentioned, some XY females with pure gonadal dysgenesis provided additional evidence mapping TDF to Yp. XY gonadal dysgenesis is a disorder in which individuals with a 46,XY karyotype are phenotypically female; they are sterile, devoid of secondary sexual characteristics, and strongly predisposed to gonadal neoplasia. Some familial cases are compat-ible with an X-linked recessive or a male limited-autosomal recessive inheritance, but most cases are sporadic. In contrast with XX males, most XY females exhibit no sex chromosome anomaly. The absence of Y-specific material has been observed in a few XY females only. This Yp deletion is not always visible with standard chromosome banding procedures. It could be shown in two such cases that the whole pseudoautosomal region of the paternal Y chromosome was absent whereas the entire pseudoautosomal region and the very distal X-specific part of Xp22.3 had been inherited from the father's X chromosome (case II of Fig. 7.5b). The Xp transfer to Yp was established by *in situ* hybridization experiments showing an Xp22.3-specific locus on Yp in both cases (Levilliers *et al.*, 1989). Thus an abnormal and terminal X–Y interchange can be found in some XY females, who appear to be the true countertype of Y(+) XX males.

The relatively low frequency of XY females with a deletion on Yp compared to the high proportion of XX males with Y-specific DNA is intriguing. Only a few cases of Yp-deleted XY females (possibly resulting from an abnormal X–Y interchange) have been reported whereas most other XY females show no Y chromosome anomaly after cytogenetic and DNA analyses. Features of Turner's syndrome have been reported for these Yp-deleted XY females but not for other XY females. A high incidence of mortality during fetal life is reported for Turner's syndrome, affecting up to 99% of the 45,X females. It is probable that Yp deletions have the same lethal effects, accounting for the distortion in the ratio of X–Y interchange XY females to Y(+) XX males.

REFINED MAPPING OF TDF AND IDENTIFICATION OF CANDIDATE GENES

It was largely admitted that mapping of TDF was achieved when Page *et al.* (1987) defined a small DNA interval apparently containing at least a part of TDF (Fig. 7.6). The proximal boundary of this interval was defined as the proximal extremity of the shortest Y chromosomal segment so far observed in an XX male (case A of Fig. 7.6). The distal boundary of the interval was defined by the distal breakpoint of the Y chromosome deletion observed in a 46,Xt(Y;22)(p11.2;q11) female with a reciprocal translocation between Y and autosome 22 (case B of Fig. 7.6). A gene encoding a zinc finger protein and called ZFY had been found in this interval and because of its probable DNA binding properties looked like a very good candidate for TDF. In addition the same gene is found on the Y chromosome of placental mammals and two copies of this gene (Zfy1 and Zfy2) are located in the Sxr region of mice. Interestingly, ZFY is closely related to an X-located gene ZFX (see Fig. 7.4 for location) that escapes X-inactivation, and both loci probably originated from a common ancestor (Table 7.1).

However, several later observations cast serious doubt on the identity of TDF and ZFY. It was first shown that ZFY and ZFX are not located on the sex chromosomes in marsupials. Furthermore expression of Zfy1 (the murine copy expressed in testis) is not detected at different developmental stages in testes of We/We mutant mice, in which germ cells cannot colonize testicular

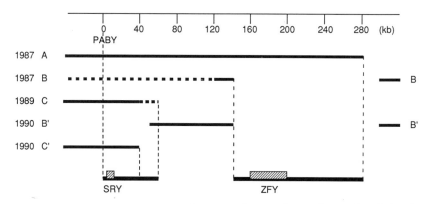

Fig. 7.6. Refined deletion mapping of TDF. A first analysis of A and B defined an interval containing ZFY. C: analysis of several XX males resulted in a new location of TDF in an interval containing SRY. B': reanalysis of B showing the double deletion. C': a more accurate definition of the TDF interval using the cases analysed in C. PABY: pseudoautosomal boundary of the Y chromosome

Table 7.1. Chromosomal assignment of cloned mammalian Y-specific genes; i.e. mapping to the sex-specific part of the Y chromosome in man and/or mouse. This table does not include the human pseudoautosomal genes, nor the processed and retransposed γ-actin and argininosuccinate synthetase pseudogenes (ACTGP2 and ASSP6). X and Y indicate the sex-specific parts (non-pseudoautosomal) of these chromosomes; Ψ corresponds to the pseudoautosomal region and Aut to an autosome. A indicates that the gene is active or presumably active and escapes X-inactivation (when X-located); a indicates that the gene is active but undergoes X-inactivation; I corresponds to an inactive pseudogene. E corresponds to an expressed sequence; +indicates presence of a related sequence with an unknown activity status; –indicates that no sequence related to that gene has been observed

Genes	Chromosomal location															
	Human				Mouse				Other eutherians				Marsupials			
	Y	X	Ψ	Aut	Y	X	Ψ	Aut	Y	X	Ψ	Aut	Y	X	Ψ	Aut
SRY	A	–	–	–	A	–	–	A	–	–	–	A	–	–	–	
TSPY	A	–	–	–	–	–	–	–	A	–	–	–				
YRRM	A	–	–	–	A	–	–	–	A	–	–	–				
UBE1	–	A or a	–	–	A	a	–	–	+	–	–	+	+	–	–	–
RPS4Y/X	A	A	–	–	A	a	–	–								
ZFY/X	A	A	–	–	A	a	–	E[b]	+	+	–	–	–	–	–	A
SMCY/X	A	A	–	–	A	A	–	–	+	+	–	–	+	+	–	–
STS	I	A	–	–	–	–	a	–					–	–	–	A
AMG	E[a]	A or a	–	–	–	–	+	–								
KAL	I	A	–	–					–	+	–	–				

[a] AMGL, the Y copy highly homologous to the X-linked AMG gene, is transcribed but cannot complement mutation of the X locus. [b] Autosomal copy of Zfx (Zfa) has arisen by retrotransposition and is transcribed.

tissue. Zfy expression thus seems associated with the presence of germ cells and has possibly a role in germ cell development, but is not required for testis differentiation. In addition Zfy1 and Zfy2 are still present and functional in XY females mutated in Tdy in which a sex reversal mutation giving rise to a fertile female phenotype co-segregates with the Y chromosome. This mutation can be complemented by a functional Sxr region. Last but not least, some human XX males which lack ZFY nevertheless harbour a small segment of the human Y chromosome of less than 60 kb called SRY (for sex-determining region) and located distal to ZFY and adjacent to the proximal pseudo-autosomal boundary (Fig. 7.6, case C). Moreover, the t(Y;22) translocation defining the ZFY interval was shown to contain a second deletion overlapping the SRY segment and extending into the pseudoautosomal region (Fig. 7.6, case B'). A gene located in the SRY interval was then identified by Goodfellow and colleagues (Sinclair *et al.*, 1990). This gene, also termed SRY, is conserved in a Y-specific manner among mammals. Its central part consists of a conserved putative DNA binding motif, similar to those present in the Mc mating-type protein of the fission yeast *Schizosaccharomyces pombe* and in the nuclear high-mobility group proteins HMG1 and HMG2, and was therefore termed the HMG box. Again, the homologous mouse gene, Sry, was found in the Sxr region (Gubbay *et al.*, 1990). But contrary to Zfy, it is deleted in the XY female mice defective for Tdy. In addition, Sry is expressed in the urogenital ridge of mouse embryos between day 11.5 and 12.5—the period during which testes begin to form.

VALIDATION OF THE CANDIDATE GENE SRY

These very strong clues were further confirmed by the finding of point mutations in human XY females demonstrating that this gene is necessary for testicular differentiation. At present a number of mutations, including small deletions, nonsense and missense point mutations, have been found in SRY. All these mutations occur in the HMG box region of the gene. A few other mutations found in familial cases and also mapping to the HMG box do not lead to a unique phenotype but can be carried by normal males or by 46,XY sex-reversed females within the same families.

In the mouse, a genomic fragment of 14 kb containing Sry was shown to be sufficient for male sex determination when introduced into XX mouse embryos (Koopman *et al*, 1991). Among 11 XX mice transgenic for Sry, three showed development of testes. In similar experiments using human SRY no evidence for testis differentiation could be observed. Although in such XX embryos SRY was transcribed between day 11.5 and 12 at a higher level than the endogenous Sry in normal XY male embryos, it appears that SRY does not function in mouse (Koopman *et al.*, 1991). Several lines of explanations could account for this difference between Sry and SRY, but the important alterations between the primary structures of both proteins within and outside the HMG domain is a very plausible one.

Although SRY/Sry is a master regulating gene it is itself subject to very stringent regulation. Factors controlling transcription of SRY/Sry and the targets of the SRY protein will be the next elements to identify in the sex determination pathway. It has been shown in gel retardation assays that the consensus DNA sequence A/TACAAT binds to recombinant SRY protein. This suggests that the SRY protein probably interacts *in vivo* with a similar DNA motif. However, it remains to be demonstrated that this DNA binding activity has a regulatory effect on transcription. Interestingly, all sporadic mutations induce a loss of DNA binding activity as shown with a number of mutated recombinant SRY proteins, whereas the mutated proteins found in familial cases showed diverse behaviours.

Though clearly sufficient for sex determination in mouse, the debate around SRY in man is not yet closed. On the one hand, the four ZFY(−) SRY(+) cases of human XX maleness described so far have either ambiguous genitalia or are only partially sex-reversed. Thus SRY may not determine the entire male phenotype that is observed in almost all cases of ZFY(+) XX males, suggesting the existence of (an)other gene(s) located in the Y chromosomal interval between the pseudoautosomal boundary and ZFY. Alternatively, in these exceptional cases the proximity of the X–Y translocation breakpoint may induce a position effect, possibly related to X-inactivation, which could decrease the level of SRY expression. On the other hand, defects of SRY have not been observed in the majority of XY females (about 85% of cases). Conversely, a small fraction of XX males (5–10%) do not harbour SRY nor other segments of the Y chromosome and are hence designated Y(−) XX males. This suggests that other genes located on X or autosomes acting either up- or downstream from SRY remain as critical as SRY in sex determination.

The peculiar location of SRY next to the pseudoautosomal boundary in man is very intriguing, but might represent an ultimate position of an evolutionary process, which SRY reached progressively or abruptly. If a more appropriate control of its expression can be achieved in such a location, this will compensate for the disadvantage of its frequent mobilization in abnormal X–Y interchanges resulting from its extreme position.

SPERMATOGENESIS AND MALE-SPECIFIC ANTIGENS

GENETIC ASPECTS

The mammalian sex chromosomes originated from a pair of autosomes which progressively diverged. It is difficult to conceive that the difference of a single locus (TDF) is sufficient to promote this drift, which was more probably favoured by the synteny of several genes involved in male production. It is therefore highly likely that the Y chromosome encodes other functions involved in male fertility. One such factor (AZF) has been mapped to the proximal part of the long arm (Yq11.23), because small cytogenetically detectable deletions of the euchromatic part of the Y chromosome were found in association with azoospermia (Fig. 7.2). Molecular approaches have allowed this mapping to be refined. Reliable correlations could be established in families where a case of male sterility (associated with a small deletion) was found along with normal male sibs (without a deletion). In addition, spermatogenesis is a process requiring the activity of numerous functions, which when impaired can all result in an AZF phenotype. Several Y-encoded genes might therefore be involved in fertility and complicate deletion mapping.

The first products encoded by Y-specific genes were two male-specific antigens: the minor histocompatibility antigen H-Y (at present defined by T cell-mediated transplantation tests) and the serologically determined male antigen SDMA. The H-Y antigen was first hypothesized to correspond to TDF. But a variant of the Sxr mutation which arose from an Sxr mouse has been observed and males with this mutation fail to express the male-specific antigen H-Y (McLaren et al., 1984). This was the first demonstration excluding H-Y as the TDF in mammals. A similar T cell-mediated transplantation test later indicated that human Y(+) XX males are also H-Y-negative. The original H-Y-positive Sxr mutation is symbolized by Sxra and the H-Y-negative variant by Sxrb. A structural analysis of the Sxrb variant has shown that the Sxr region of Sxra has undergone a deletion (ΔSxrb) through unequal recombination between Zfy1 and Zfy2 (Fig. 7.7). It was shown that ΔSxrb also contains Sdma—the locus controlling expression of the serologically deter-mined male antigen (Fig. 7.7).

Adult XX Sxra mice have sterile testes, probably as a consequence of the perinatal loss of germ cells. It is supposed that expression of a second X chromosome active in germ cells impairs spermatogenesis. On the contrary, all stages of spermatogenesis can be observed in adult XO Sxra mice (Burgoyne et al., 1986). They remain nonetheless sterile because the later stages are strongly underrepresented. In XO Sxrb there is an apparent block of spermatogenesis around the spermatogonial stage and the more mature stages cannot be observed. Sxrb mice have thus also lost genetic information required for spermatogenesis (Burgoyne et al., 1986). This gene has been called Spy, and encodes a factor involved in the proliferation and survival of A spermatogonia. Based on mapping, the H-Y antigen might be a candidate for this spermatogenesis factor. Similar conclusions can be reached for the human Y chromosome where the HY locus has been mapped to the proximal long arm as the fertility factor AZF (Fig. 7.2).

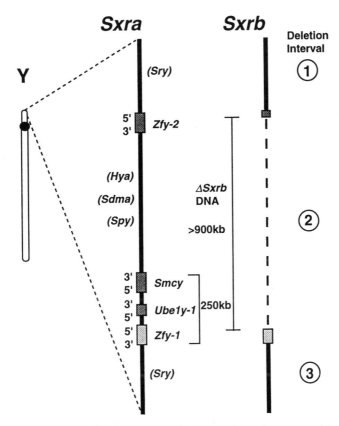

Fig. 7.7. A physical map of the sex-reversed region of the short arm of the mouse Y chromosome. Sxrb is a deletion mutation of the original Sxra. The following genes map to this deletion interval: Hya, which codes or controls the expression of the minor histocompatibility antigen H-Y, as measured by cellular T cell assays, Sdma, a gene controlling the expression of the serologically defined male antigen (which may or may not be the same as H-Y), and Spy, a gene controlling spermatogonial proliferation and/or survival (Burgoyne *et al.*, 1986). The size of the Sxrb deletion is greater than 900 kb. The relative position of loci in parentheses is not known. Sry could map to either interval 1 or 3. Reproduced from Agulnik *et al.* (1994), *Hum. Mol. Genet.*, 3, 873–878, by permission of Oxford University Press

MOLECULAR ASPECTS

At the molecular level several genes possibly involved in spermatogenesis have been identified both in man and mouse, but in no case is there definite proof that these genes correspond to HY/Hya or AZF. In man, a gene called TSPY has been found on the proximal short arm (Fig. 7.2) and encodes a testis-specific transcript conserved in bovine DNA but apparently not found in mouse. Recently a family of candidate sequences for AZF, called YRRM, has been identified (Ma *et al.*, 1993). Members of this YRRM family map to a region of Yq11.23 containing microdeletions in some azoospermic males, are expressed in adult testis, and their predicted protein products share homology with known RNA binding proteins. YRRM sequences are conserved in a Y-specific manner among mammals, including the mouse. But direct evidence that absence of products of this gene family impairs spermatogenesis is not yet available.

A murine testis-specific transcript, termed Ube1y-1, expressed from the ΔSxrb region (Fig. 7.7) has been cloned (Mitchell *et al.*, 1991). Ube1y-1 has a homologue on the X chromosome (Ube1x) and encodes a protein homologous to the human ubiquitin activating enzyme E1 (UBE1) also mapping to Xp11.2. Escape from X-inactivation has been reported in humans but remains controversial, whereas Ube1x is inactivated in mouse (Kay *et al.*, 1991). The gene is highly conserved on both sex chromosomes in mammals but is only found on the X chromosome in higher primates, including man. Mapping, pattern of expression and biochemical properties exclude identity with HY. It has also been observed that UBE1 is able to correct a temperature-sensitive mutation tsA1S9 of mouse L cells. This mutation has been shown to affect nuclear DNA replication. Other ts mutants with a thermolabile E1 enzyme are arrested in the cell cycle or DNA replication. The involvement of UBE1 in regulating the mitotic cell cycle suggests that UBE1/Ube1y-1 might exert a similar role in meiosis and could thus be involved in spermatogenesis.

Another mouse Y-specific gene, Smcy, mapping to the DSxrb region (Fig. 7.7) has been cloned recently. It is ubiquitously expressed in male tissues from the initial state of embryonic development. Its human homologue maps to the deletion interval containing HY on Yq11.23. It has also an X-specific homologue on both murine and human X chromosomes and escapes X-inactivation in both species. This is the first example of a gene escaping X-inactivation in the mouse.

THE CASE OF THE TURNER GENE(S)

Turner's syndrome is a complex phenotype normally associated with a 45,X karyotype. Turner females are characterized by short stature, gonadal dysgenesis, neonatal lymphoedema, webbed neck, etc. As already mentioned, most cases do not come to term but undergo spontaneous abortion. It has been proposed that the Turner phenotype results from the presence of only one, rather than two active copies ('haploinsufficiency') of one or more genes shared by the X and Y chromosomes; these are termed Turner gene(s). Because karyotypic anomalies indicated that the Turner genes are located on Xp, it was proposed that such genes could be pseudoautosomal. However, X–Y interchange XX females with two copies of the pseudoautosomal region also show some signs of Turner's syndrome, which should thus be associated with the deletion found in Yp. Among these signs one finds a marked neonatal lymphoedema of the extremities but no short stature. Genes affecting size have been tentatively mapped to Yq (GCY in Fig. 7.2) and to the pseudoautosomal region and are probably present in X–Y interchange XX females as in normal males. But the 'oedema' gene(s) should be encoded by X and Y-specific loci. The Y-located 'oedema' gene(s) should map within the Yp deletion of X–Y interchange XY females.

Since ZFX maps to Xp21 and escapes X-inactivation, it meets the criteria of the X-located Turner gene. However, the 46,Xt(Y;22) translocation described by Page et al. is not associated with a Turner phenotype (Fisher et al., 1990) despite its ZFY deletion. Furthermore, this translocation is the key rearrangement for the mapping of another Turner candidate gene (Fisher et al., 1990). The gene encoding protein S4 from the small ribosomal subunit, RPS4Y, has been located to a region distal and adjacent to ZFY. Interestingly, this gene has an X counterpart, RPS4X, mapping to Xq13.1 and escaping inactivation—a prerequisite for a Turner gene. Y Turner genes were proposed to map to interval 1 or 2 on the basis of the deletion observed in X–Y interchange XY females (Levilliers et al., 1989). Interval 2 is mainly constituted of a segment of DNA translocated from Xq21 to Yp during recent human evolution (Foote et al., 1992) (Fig. 7.4: regions indicated by arrow 1). Since patients with deletions of Xq21 have no Turner phenotype, Fisher et al. infer that a Turner gene cannot map to interval 2 of Yp. Conversely, RPS4Y maps to the undeleted part of interval 1 in the non-Turner Yp deletion. On the basis of these mapping arguments and escape from X inactivation it is concluded that haploinsufficiency of RPS4Y/RPS4X contributes to the Turner phenotype. But involvement of Xq13 in Turner's syndrome remains to be documented. In addition, the isochromosome of the long arm of the X chromosome, i(Xq), is the most common structural aberration found in patients with Turner's syndrome. The presence of three copies of Xq in these patients argues against the implication of Xq deletions in Turner's syndrome.

Turner phenotype is not known in mouse and XO mice are fertile to some extent. These differences might be due to differences in X chromosome inactivation since most of the genes known to escape X-inactivation in man (RPS4X, STS, ZFY, UBE1) are inactivated in mouse (Table 7.1), SMCX/Smcx being the first exception to this general observation.

It thus appears that several X-linked genes have conserved homologues on the Y chromosome. They are listed in Table 7.1 and their X and Y chromosomal location is shown in Fig. 7.4. One characteristic of Y-located genes with X/Y homology is that, in humans at least, their X homologues escape X-inactivation, thus maintaining gene dosage between males and females. Conversely, one could also expect that genes that escape X-inactivation have active Y homologues. However, STSP and KALP, the Y homologues of STS (steroid sulphatase) and KAL (Kallmann's syndrome), respectively, no longer encode functional proteins. AMGL, the homologue for AMG (amelogenin), is still transcribed but cannot complement AMG mutations. Furthermore, though UBE1 escapes X-inactivation, no Y homologue has been found in man. Most of these genes have also evolved independently from one another and diversely in different species as reflected by the differences observed among the mammals (Table 7.1).

CONCLUSION

Molecular mapping of the human Y chromosome has been possible because most anomalies analysed were consistent with a single order of loci in chromosomes from different individuals. But as long as new rearrangements of the Y-specific part do not impair its essential functions, it should be possible to observe variants in the order of the different loci. In some rare instances of sex reversals, DNA analysis has suggested the existence of alternative orders. A cytogenic variant with a pericentric inversion has also been observed in the population of western India. Conversely, the localization of Y-specific genes is dramatically different in man and mouse. In addition, synteny is not conserved for all genes, indicating that Y-specific expression might be dispensable in some instances. This suggests that the mammalian Y chromosome is still actively evolving and that some of the functions encoded by this chromosome probably do not require to be expressed in a sex-specific manner but could instead be localized on autosomes.

REFERENCES

(Original papers have mainly been limited to recent papers on mapping or identification of Y-specific genes; older references can be found therein or in the reviews cited below.)   ‹

Agulnik AI, Mitchell MJ, Lerner JL, Woods DR and Bishop CE (1994) A mouse Y chromosome gene encoded by a region essential for spermatogenesis and expression of male-specific minor histocompatibility antigens. *Hum. Mol. Genet.*, **6**, 873–878.
Burgoyne PS, Levy ER and McLaren A (1986) Spermatogenic failure in male mice lacking H-Y antigen. *Nature*, **320**, 170–172.
Fisher EMC, Beer-Romero P, Brown LG et al. (1990) Homologous ribosomal protein genes on the human X and Y chromosomes: escape from X inactivation and possible implications for Turner syndrome. *Cell*, **63**, 1205–1218.
Foote S, Vollrath D, Hilton A and Page DC (1992) The human Y chromosome: overlapping DNA clones spanning the euchromatic region. *Science*, **258**, 60–66.
Gubbay J, Collignon J, Koopman P et al. (1990) A gene mapping to the sex-determining region of the mouse Y chromosome is a member of a novel family of embryonically expressed genes. *Nature*, **346**, 245–250.
Guellaën G, Casanova M, Bishop C et al. (1984) Human XX males with single-copy Y fragments. *Nature*, **307**, 172–173.

Kay GF, Ashworth A, Penny GD *et al.* (1991) A candidate spermatogenesis gene on the mouse Y chromosome is homologous to ubiquitin-activating enzyme E1. *Nature*, **354**, 486–489.

Koopman P, Gubbay J, Vivian N, Goodfellow P and Lovell-Badge R (1991) Male development of chromosomally female mice transgenic for Sry. *Nature*, **351**, 117–121.

Ma K, Inglis JD, Sharkey A *et al.* (1993) A Y chomosome gene family with RNA-binding protein homology: candidates for azoospermia factor AZF controlling human spermatogenesis. *Cell*, **75**, 1287–1295.

McLaren A, Simpson E, Tomonari K, Chandler P and Hogg H (1984) Male sexual differentiation in mice lacking H-Y antigen. *Nature*, **312**, 552–555.

Mitchell JM, Woods DR, Tucker PK, Opp JS and Bishop CE (1991) Homology of a candidate spermatogenic gene from the mouse Y chromosome to the ubiquitin activating enzyme E1. *Nature*, **354**, 483–486.

Page DC, Mosher R, Simpson EM *et al.* (1987) The sex-determining region of the human Y chromosome encodes a finger protein. *Cell*, **51**, 1091–1104.

Petit C, de la Chapelle A, Levilliers J *et al.* (1987) An abnormal terminal X–Y interchange accounts for most but not all cases of human XX maleness. *Cell*, **49**, 595–602.

Sinclair AH, Berta P, Palmer MS *et al.* (1990) A gene from the human sex-determining region encodes a protein with homology to a conserved DNA-binding motif. *Nature*, **346**, 240–244.

FURTHER READING

REVIEWS

Human sexual anomalies

Simpson JL (1982) Abnormal sexual differentiation in humans. *Annu. Rev. Genet.*, **16**, 193–224.

Mammalian sex determination

Eicher EM and Washburn LL (1986) Genetic control of primary sex determination in mice. *Annu. Rev. Genet.*, **20**, 327–360.

Goodfellow PN and Lovell-Badge R (1993) SRY and sex determination in mammals. *Annu. Rev. Genet.*, **27**, 71–92.

Hawkins JR (1993) The SRY gene. *Trends Endocrinol. Metab.*, **4**, 328–332.

HUMAN Y CHROMOSOME

Rappold G (1993) The pseudoautosomal regions of the human sex chromosomes. *Hum. Genet.*, **92**, 315–324.

Wolf U, Schempp W and Scherer G (1992) Molecular biology of the human Y chromosome. *Rev. Physiol. Biochem. Pharmacol.*, **121**, 147–213.

BOOKS

SEX DETERMINATION

Classical aspects
Austin CR and Edwards RG (eds) (1981) *Mechanisms of Sex Differentiation in Animals and Man*. London: Academic Press.

Molecular aspects
Wachtel SS (ed.) (1994) *Molecular Genetics of Sex Determination*. London: Academic Press.

8 Identification and Characterization of the Neurofibromatosis 1 Gene

DAVID VISKOCHIL

The successful cloning of several genes associated with inherited human diseases has established the utility of the mapping approach for isolating genes that encode proteins of unknown structure or function. For example, *deletion mapping* studies led to cloning of the genes bearing disease-causing mutations in Duchenne muscular dystrophy, chronic granulomatous disease and retinoblastoma. *Genetic linkage* studies led to the eventual identification of the genes for heritable conditions such as cystic fibrosis, familial adenomatous polyposis and, the topic of this chapter, neurofibromatosis type 1. The rationale for disease gene cloning by the mapping approach derives from the expectation that if a gene can be isolated, determination of its DNA sequence may yield clues to the function of its encoded peptide and thereby lead to a better understanding of the pathophysiology of the disease.

The cloning of disease-causing genes, when a pattern of inheritance is evident, involves a number of initial steps including definition of a phenotype, ascertainment of affected and unaffected individuals within a large pedigree, and mapping of the disease locus through genetic linkage. Mapping by genetic linkage in family studies utilizes DNA-based genetic markers at polymorphic loci to establish a segregation pattern matching that of the disease phenotype. Evidence of genetic linkage to previously mapped markers provides an approximate chromosomal location for the unknown gene. More precise mapping through additional, closely spaced genetic markers, often with the aid of microscopically visible chromosomal rearrangements, can define a relatively short genomic segment as the disease locus. Isolation of overlapping genomic clones that together form a continuous stretch of DNA sequence, or a 'contig', spanning this segment yields DNA probes that may identify transcribed genes from the locus. The disease gene is identified through mutational analysis of these 'candidate' genes in the DNA of affected individuals. Using the mapping approach the neurofibromatosis 1 gene was cloned approximately 3 years after establishing genetic linkage to chromosome 17.

Molecular Genetics of Human Inherited Disease. Edited by D.J. Shaw
Published 1995 by John Wiley & Sons Ltd

NEUROFIBROMATOSIS 1: CLINICAL ASPECTS

Neurofibromatosis 1 (NF1), the von Recklinghausen or peripheral form of neurofibromatosis, was an early target for the mapping approach. The neurofibromatoses are a heterogeneous set of conditions, each characterized by nervous system tumours, of which NF1 is the most common. This chapter is devoted specifically to the isolation and characterization of the gene causing NF 1 (see Table 8.1). NF1 is one of the most prevalent of human genetic disorders, as it affects about 1 in 3500 individuals worldwide. It is characterized primarily by tumours associated with the peripheral nervous system and by café-au-lait spots (CLS) on the skin. Although NF1 is inherited in an autosomal dominant fashion, approximately half of the cases diagnosed are sporadic occurences—new mutations in the germ-line, where there is no prior family history of the disease. Although NF1 is fully penetrant, its clinical expression is highly variable even among members of the same family. This characteristic of NF1 necessitated development of diagnostic criteria which, in addition to assisting with the diagnosis, provided objective clinical ascertainment of affected individuals in the families enrolled in genetic linkage studies. The features of NF1 that recommended it for the mapping approach included the following: high incidence in the worldwide population, full penetrance, and autosomal dominant pattern of inheritance. Moreover, the absence of biochemical clues to the disease process left little hope for the development of rational therapeutic intervention. Cloning of the NF1 gene has provided the means for determining DNA and translated amino acid sequence, information useful in predicting structure and/or function of the NF1 protein. Such data are expected to provide a basis for a more rational approach to the development of effective treatment.

NF1 is a clinical diagnosis whereby a defined set of clinical criteria (see Table 8.2) need to be satisfied (Stumpf *et al.*, 1988). The features typically seen in this condition include multiple CLS, neurofibromas, freckling in the axillae and groin, optic nerve gliomas, Lisch nodules and specific osseous lesions. NF1 is present in an individual if two or more of these criteria are satisfied. Generally the skin manifestations have little clinical significance, and approximately two-thirds of affected individuals lead normal lives. The indi-

Table 8.1. Features of neurofibromatosis 1

Autosomal dominant
Incidence of 1 in 3500
Half of cases have no family history of NF1
Clinical phenotype characterized by
  Multiple subcutaneous nodules (neurofibromas)
  Hyperpigmented patches of skin (café-au-lait spots)
Relevance to cancer

Table 8.2. Clinical criteria for the diagnosis of NF1 (Stumpf *et al.*, 1988)

Six or more café au lait macules over 5 mm in greatest diameter in pre-pubertal individuals and over 15 mm in greatest diameter in post-pubertal individuals
Two or more neurofibromas of any type or one plexiform neurofibroma
Freckling in the axillary of inguinal regions
Optic glioma
Two or more Lisch nodules (iris harmartomas)
A distinctive osseous lesion such as sphenoid dysplasia or thinning of long bone cortex with or without pseudoarthrosis
A first-degree relative (parent, sibling or offspring) with NF1 by the above criteria

---

The diagnostic criteria are met in an individual if two or more of the features listed are present.

viduals who suffer medically significant problems show any number of the following clinical features: paraspinal neurofibromas, renal artery hamartomas, phaeochromocytomas, neurofibrosarcomas, leukaemia of myelogenous type, hypertension, learning disabilities, seizures, scoliosis and visual disturbances. The clinical variability of NF1 is striking; the development of medical complications is unpredictable even within a family whose affected members presumably carry an identical mutation. Even though NF1 is a progressive condition, whereby affected individuals develop increasing numbers of clinical features as they age, most people have a near-normal life expectancy.

The natural history of NF1 follows a pattern of age-related penetrance for a number of clinical manifestations. Cutaneous neurofibromas rarely become apparent before adolescence, at which time an increase in the number and size of lesions is common. Generally, by the age of 30 numerous neurofibromas appear, mainly over the trunk area. In contrast to the cutaneous and discrete neurofibromas, plexiform neurofibromas are sometimes present at birth. Congenital neurofibromas are usually of the plexiform type even though the involvement with surrounding tissue may not be apparent. These tumours often present difficulties for clinical management because they can arise almost anywhere in the body; also, because they typically involve a number of tissues in an interdigitating fashion, such tumours are difficult to resect surgically. CLS may also present congenitally, but more typically they arise during infancy and early childhood, so that by 6 years of age the number and size of CLS are reliable diagnostic criteria for NF1. Iris hamartomas or Lisch nodules tend to appear in late childhood through early adulthood and, in recent surveys, approximately 95% of individuals with NF1 have this medically insignificant marker that is rare in the unaffected population. Even though NF1 is fully penetrant, the variability of clinical expression and progressive nature of this condition make long-term medical surveillance imperative; annual visits to a clinic where medical specialists are familiar with NF1 are recommended.

The pathophysiology of NF1 is not understood, and the underlying biochemical defect in NF1 is not known. It is assumed that aberrant growth of neural crest-derived tissue is, in some way, responsible for many manifestations of NF1. Neurofibromas, the hallmark of this condition, are benign tumours arising from the peripheral nervous system. They are intimately associated with myelinated nerve axons and are composed of extracellular matrix (laminin, collagen and proteoglycan), Schwann cells, fibroblasts, mast cells, endothelial cells and perineural cells. As Schwann cell hypercellularity is a consistent finding in neurofibromas, it is generally accepted that proliferating Schwann cells are responsible for growth of a neurofibroma. Aberrant growth and/or migration of melanocytes, another neural crest-derived cell type, lead to the CLS associated with NF1. The cellular mechanisms specifying neural crest-derived tissue as the predominant cell types that demonstrate most of the clinical manifestations of NF1 are unknown.

The increased frequency of malignant tumours in patients with NF1 is well documented and is certainly one of the most worrisome aspects of this condition. Counselling NF1 patients about the likelihood of cancer is made difficult by the unpredictability of clinical expression in any one individual. Although the epidemiological studies regarding NF1 and its link to malignancy are naturally biased in the patient selection process, there is clearly an increase in malignancy and a shift of the age-specific incidence curve for cancer (Lefkowitz et al., 1990). The most common malignancies associated with NF1 are neurofibrosarcoma, rhabdomyosarcoma, phaeochromocytoma, astrocytoma, duodenal carcinoid and chronic juvenile myelogenous laeukemia, all of which are difficult to treat medically. Like many other genetic conditions, the unpredictability of the severe manifestations of NF1 create a 'time-bomb' apprehension in patients, and the anxiety provoked is a source of serious psychological stress. Cloning of the NF1 gene holds promise for the development of effective treatment regimens for NF1 patients; moreover, characterization of mutations in the NF1 gene may lead to correlations between genotypes and phenotypes that could help clinicians predict the natural history of NF1 on an individual basis.

ISOLATION OF THE NF1 GENE

GENETIC LINKAGE MAPPING

The first step toward cloning the NF1 gene by the mapping approach involved linking the disease-causing allele to previously mapped DNA markers. Critical to any genetic linkage study is ascertainment of affected and unaffected individuals in families that carry the trait through multiple generations. The heterogeneity of NF1, even within families, made it imperative to apply strict clinical criteria to ascertain persons with *NF1* mutations. Using similar diagnostic criteria to those shown in Table 8.2, two research groups were able to map *NF1* to the long arm of chromosome 17 (Barker *et al.*, 1987; Seizinger *et al.*, 1987). Barker *et al.* demonstrated a maximum LOD score of 4.2 for linkage with a centromeric DNA marker, and Seizinger *et al.* found linkage to the nerve growth factor receptor with a LOD score of 4.4. Refined linkage analysis localized the *NF1* gene to a relatively closer linkage with the centromeric probe pHHH202 (D17S33) and mapped the *NF1* locus within band q11.2 of chromosome 17.

Knowing the map position of a disease locus is merely the first step in cloning the gene. Markers showing 0% genetic recombination in linkage studies might be physically separated by millions of base pairs. A major achievement in cloning the *NF1* gene was the transition from a genetic linkage map to a physical map of the *NF1* locus. Two primary strategies used to accomplish this feat were the identificaiton of disease-causing chromosomal rearrangements, and the selection of DNA sequences, or probes, physically mapping to the disease locus. In order to saturate the *NF1* locus with probes, chromosome 17-specific genomic clones were generated to pepper the target region defined by linkage studies (O'Connell *et al.*, 1989a). This was accomplished by taking advantage of the location of the thymidine kinase locus on chromosome 17q21–22; this marker allows the development of rodent × human somatic cell hybrid clones by culture in a selective growth medium (HAT). Probes derived from chromosome 17 DNA were used to identify cell lines containing segments of chromosome 17 material encompassing the *NF1* locus. Genomic DNA from such cell hybrids was cut with restriction enzymes and cloned into cosmid vector 'libraries'. Clones containing human DNA sequence were selected by probing the 'library' with total human DNA and each clone represented approximately 30 kb of DNA derived from human chromosome 17. The task then was to physically and/or genetically map each of the cosmid clones back to the region of chromosome 17 encompassing the *NF1* locus.

Concurrent with the generation of random genomic clones from chromosome 17 was the fortuitous finding of balanced translocations involving chromosome 17q11.2 in two unrelated NF1 patients. These cytogenetically visible translocations provided a breakthrough for the physical mapping of the *NF1* locus. The first chromosomal rearrangement, 46XX,t(1;17) (p34.3;q11.2), as shown in Fig. 8.1(a), was ascertained in a woman with NF1 who had suffered recurrent miscarriages. This translocation breakpoint involved the *NF1* region that had been previously defined by genetic linkage analysis. The second rearrangement, 46XX,t(17;22)(q11.2;q11.2), as shown in Fig. 8.1(b), was identified in the mother of a child with multiple congenital anomalies and an unbalanced chromosomal translocation. The mother fulfilled the clinical criteria for NF1, and her chromosomal analysis demonstrated a balanced translocation, with the translocation breakpoint involving

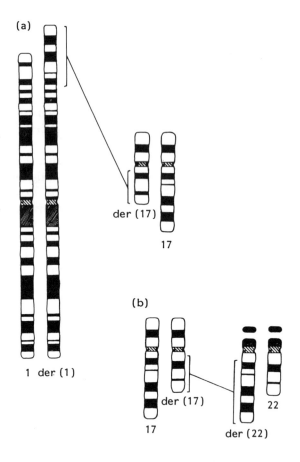

Fig. 8.1. Two balanced translocations involving the centromeric region of chromosome 17 in unrelated NF1 patients. (a) Balanced translocation involving 17q11.2 and 1p34.3. (b) Balanced translocation involving 17q11.2 and 22q11.2

the region on 17q encompassing the *NF1* locus. Presumably both translocation breakpoints interrupted the normal expression of the putative *NF1* gene and caused the *NF1* phenotype. Thus, precise mapping of the breakpoints with clones derived from chromosome 17 would provide probes for the identification and isolation of the *NF1* gene.

PHYSICAL MAPPING OF THE NF1 LOCUS

The technique of pulsed-field gel electrophoresis (PFGE) was used to make a physical map of the cloned DNA sequences around the *NF1* locus. PFGE separates large fragments of DNA, up to several megabases in size, and enables one to map large regions of chromosomes by hybridization of probes to Southern blots of DNA digested with restriction enzymes that cut very infrequently, and separated by PFGE. As shown in Fig. 8.2, two probes closely flanking the NF1 locus (17L1A and 11-1F10) hybridized to the same 600 kb fragment of DNA generated by an NruI digest. Further PFGE studies showed that both probes, although separate from each other, were able to detect abnormally sized NruI fragments in DNA from the two NF1 patients with balanced chromosome translocations. This result showed that both translocation breakpoints were located within the 600 kb NruI fragment. The development of human–rodent hybrid cell lines in which the translocation chromosomes were isolated provided material for a panel of cell lines useful for further mapping of the NF1 region (Fig. 8.3). Southern blot analysis using the hybrid cell lines and the two probes mentioned above showed that the breakpoints in the NF1 chromosomal translocations were physically separate, with the t(17;22) breakpoint nearer to the telomere than the t(1;17).

The region between the breakpoints now became the focus of efforts to obtain further cloned DNA, since this interval was the best candidate for containing the NF1 gene itself. Although a number of useful genomic DNA sequences were isolated, a chance finding provided the most direct access to the NF1 region. A mouse myeloid leukaemia gene associated with an

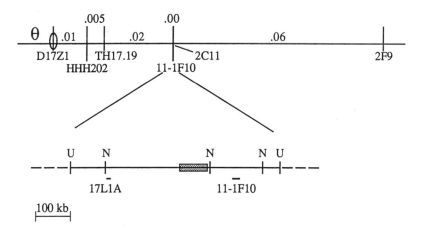

Fig. 8.2. Transition from genetic linkage mapping to physical mapping of the NF1 locus. The upper line represents the centromeric region of chromosome 17 and the genetic map position of randomly acquired probes. The lower line represents an NruI, (U), fragment from the 2C11 map site and the physical mapping of two probes, 17L1A (Fountain et al., 1989) and 11-1F10 (O'Connell et al., 1989b), to adjacent NotI, (N), fragments. The shaded box on the physical map represents an approximate location of the t(1;17) and t(17;22) breakpoints; θ represents recombination estimates

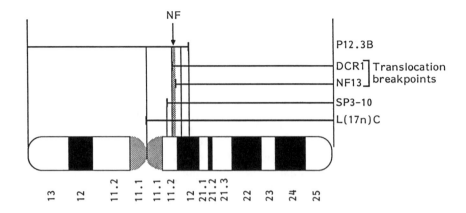

Fig. 8.3. Somatic cell hybrid panel used to physically map the NF1 region. The NF1 region is labelled at band 11.2 on chromosome 17 and the portions of chromosome 17 present in each hybrid are shown above the diagram. L(17n)C contains all of 17q; SP3-10 is a hybrid cell line containing the derivative chromosome with the 15 centromere from a 15;17 translocation; NF13 contains the derivative chromosome with the 22 centromere from the NF1 patient with the t(17;22); DCR1 contains the derivative chromosome with the 1 centromere from the NF1 patient with the t(1;17); P12.3B contains the derivative chromosome with the 17 centromere from a patient with acute promyelocytic leukaemia and a balanced translocation involving chromosomes 15 and 17. The shaded region between NF13 and DCR1 represents the NF1 region (not drawn to scale)

ecotropic virus insertion site, *Evi-2*, had been mapped to mouse chromosome 11. Because this chromosome shows extensive homology to human chromosome 17, Buchberg and Copeland (National Cancer Institute, Maryland, USA) collaborated with O'Connell and White (Howard Hughes Medical Institute, Salt Lake City, USA) to map the human homologue for *Evi-2*. The association of this gene with myeloid leukaemia in mice led to the idea that the human version might map to a t(15;17) breakpoint found in human acute promyelocytic leukaemia; this breakpoint was represented in the somatic cell hybrid panel used to map the *NF1* region (Fig. 8.3). However, linkage and physical mapping studies showed that rather than mapping to the t(15;17) breakpoint region, the human version of *Evi-2* was located between the two *NF1*-associated breakpoints (probe pHU39.3, Fig. 8.4a). Mapping this gene between the two breakpoints was a significant development in the construction of a high-resolution physical map of the *NF1* region (Fig. 8.4b). The construction of this map, in which probes shown to be genetically linked to *NF1*, and the disease-associated chromosome breakpoints, were integrated into the physical map of this part of the chromosome, transformed the perception of the *NF1* locus; it could then be viewed as a physical entity, with

(a)

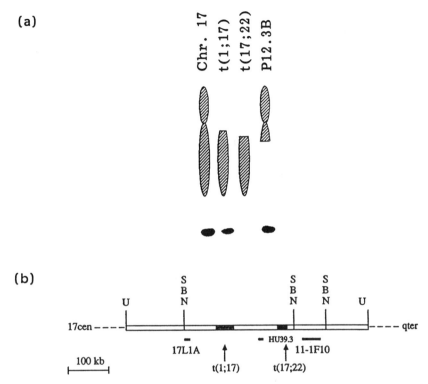

(b)

Fig. 8.4. Identification of a clone that maps between the translocation breakpoints. (a) Southern analysis of *Bg*lII-digested DNA from somatic cell hybrids containing human chromosome 17; DCR1, t(1;17); NF13, t(17;22); and P12.3B, t(15;17). The autoradiograph demonstrates hybridization of pHU39.3 to a 10 kb fragment in all cell lines except NF13. (b) Pulsed-field gel electrophoresis map for the 600 kb *Nru*I fragment containing the *NF1* translocation breakpoints. Cleavage sites: B, *Bss*HII; N, *Not*I; S, *Sac*II; U, *Nru*I. Cross-hatching represents the approximate locations of the translocation breakpoints. The map positions of probes 17L1A, HU39.3 and 11-F10 are depicted by solid bars

its own restriction map 'signature' to distinguish it from the rest of the genome. It also demonstrated the efficacy of the mapping approach in the isolation of human disease genes.

Establishing the location of the human *Evi-2* gene between the translocation breakpoints was a critical step in the isolation of the *NF1* gene. The probe could be used to identify clones that would likely contain exons for the gene. Two cosmids (EVI20 and EVI36) that mapped to the breakpoint region were identified. 'Walking' from these identified two more cosmids, and a contiguous region of about 80 kb was thus defined (Fig. 8.5). Southern blot analysis with probes from the cosmids defined the positions of the two breakpoints as 50 kb apart. Analysis by PFGE showed that both translocation breakpoints

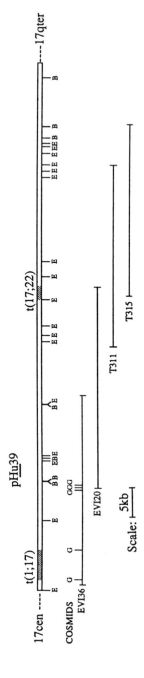

Fig. 8.5. Cosmid contig and restriction enzyme map of the *NF1* region as defined by the translocation breakpoints. E, *EcoRI*; B, *BamHI*; G, *BgIII*. The translocation breakpoint regions (shaded) are separated by approximately 55 kb. The map position of pHU39.3 as determined by Southern blot analysis is shown

were located in the middle third of a 350 kb *Not*I restriction fragment. The significance of mapping sequences with respect to sites for the enzyme *Not*I (GCGGCCGC) stems from the observation that CpG dinucleotides, when unmethylated and hence able to be cut by *Not*I, are often located in clusters associated with active genes, known as CpG or HTF islands. Under this assumption, genomic DNA containing *Not*I sites should be an enriched source of coding sequences.

However, the assumption that chromosome translocation breakpoints associated with a clinical phenotype would interrupt the expression of a critical gene, directed the search for *NF1* exons away from the *Not*I sites and toward the breakpoint region. Since the cosmids spanned this region (Fig. 8.5) it was reasonable to assume that exons of the *NF1* gene would be represented in the cosmids. Therefore, probes from the cosmids were used to identify human genes from various cDNA libraries.

IDENTIFICATION OF GENES AT THE *NF1* LOCUS

The first of three cDNAs (see Fig. 8.6) initially identified by probes mapping to the region between the breakpoints was *EVI2*, a 1.6 kb transcript found in normal human brain and peripheral blood mononuclear cells (Cawthon *et al.*, 1990a). The entire gene is represented in two exons spanning less than 10 kb of genomic DNA, and the coding region is completely contained within one exon. *EVI2* encodes a protein of 232 amino acids that may function as a cell surface receptor; however, no functional studies have been carried out on this protein. Initially this was an obvious candidate gene because of its homology with the murine locus *Evi-2*, which was known to be associated with neoplasia in the mouse. However, no mutations were found which could distinguish *EVI2* in NF1 individuals from *EVI2* in unaffected people.

A second gene identified from the breakpoint region was *RC1* (Cawthon *et al.*, 1991). This gene also has a murine homologue mapping to the *Evi-2* locus; it lies approximately 3 kb downstream in the mouse genome from the gene previously designated *Evi-2*. Because a number of viral integration sites have been found to interrupt genomic DNA encompassing both mouse genes, one

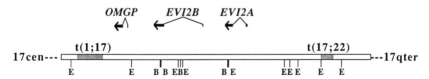

Fig. 8.6. Restriction enzyme map of the *NF1* region. E, *Eco*RI; B, *Bam*HI; G, *Bgl*II. Three genes lying between the translocation breakpoints are shown above the restriction enzyme map. The direction of transcription is shown by an arrow (5′→3′), exon sequence is shown by bold lines, and intron sequence is shown by thin lines connecting the exons

cannot predict which one causes retrovirus-induced myeloid tumours in mice; therefore, the adopted nomenclature for the murine locus is derived from the *ecotropic viral insertion* cluster site number 2 (site 1 maps to a different murine chromosome), genes *A* and *B*. This scheme gives no preference to which gene may be oncogenic in mice and the human homologues were renamed accordingly: *EVI2A* replaced *EVI2*, and *EVI2B* replaced *RC1*. *EVI2B* serves as a template for a 2.1 kb transcript found exclusively in bone marrow, peripheral blood mononuclear cells and fibroblasts. The *EVI2B* locus spans approximately 10 kb of genomic DNA that includes 57 bp of 5' non-coding exon, 8 kb of intronic sequence, and a 2078 bp exon that carries the entire open reading frame encoding 448 amino acids. *EVI2B* is in the same transcriptional orientation as *EVI2A*; its 5' exon lies approximately 4 kb downstream from the end of the 3' exon of *EVI2A* (see Fig. 8.6). Even though the predicted amino acid structure is proline-rich and has features common to transmembrane proteins, the function of *EVI2B* has not been determined. Furthermore, multiple screening techniques did not detect mutations in this gene among 44 individuals with NF1.

OMGP, a third gene from this region, was identified by screening cDNA libraries with an evolutionarily conserved DNA sequence that mapped between the two translocation breakpoints and within 5–10 kb of the t(1;17) translocation breakpoint (Viskochil *et al.*, 1991). Even though this cDNA was identified with a conserved sequence, no rodent homologue for *OMGP* has been reported. In humans, *OMGP* encodes oligodendrocyte myelin glycoprotein, a component of the myelin sheath, that is elaborated by the oligodendrocytes of the central nervous system. *OMGP* spans at least 2.5 kb of genomic DNA; it is transcribed as a 1.8 kb message that encodes a 421 amino acid protein. The 5' end of the gene has not been established, nor have transcriptional promoter regions in the genomic DNA sequence upstream from the most 5' end been identified. *OMGP* lies 1–4 kb distal to the t(1;17) breakpoint and approximately 5 kb proximal to *EVI2B*. Its genomic structure is similar to both *EVI2A* and *EVI2B* in that it has only two exons, one of which contains the entire open reading frame. It is comprised of a 77 bp non-coding 5' exon, an 814 bp intron, and a 1688 bp coding 3' exon, in the same transcriptional orientation as *EVI2A* and *EVI2B*. *OMGP* seemed an especially promising candidate for the *NF1* gene because of its proximity to the t(1;17) translocation breakpoint and its postulated role as a cell adhesion molecule in oligodendrocytes. However, again no mutations in *OMGP* were found in NF1 individuals.

TOWARD THE IDENTIFICATION OF THE *NF1* GENE

The importance of mutation analysis in the identification of disease genes cannot be overstated. As already described, three likely candidate genes mapping to the *NF1* locus, as defined by two translocation breakpoints, did not have any base pair mutations specific to NF1 patients. Concurrent with the search for single base pair mutations in candidate genes, larger-scale genomic rearrangements were being sought among a panel of 120 unrelated NF1 patients. Southern blot analysis, using restriction enzymes that cut human DNA into fragments in the 100 bp to 10 kb size range separated by conventional agarose gel electrophoresis, was performed to identify me-dium-sized DNA rearrangements. To identify larger rearrangements in the 15–300 kb size range, PFGE analysis was performed with infrequently cutting restriction enzymes (NotI, SacII and BssHII). The probes used were derived from the cosmids that spanned the translocation breakpoints (Fig. 8.5). In either analysis, the presence of restriction fragments in NF1 samples that were absent from normal control samples would strongly suggest the presence of a DNA rearrangement causing the NF1 phenotype. Using PFGE conditions that favoured the detection of larger rearrangements (50–300 kb), no muta-tions were found in 80 patients. Likewise no mutations were found in 120 patients using conventional gel analysis. However, when the PFGE condi-tions were modified to increase resolution in the 15–100 kb range, and better probes were developed for the conventional gel analysis, deletions within the translocation breakpoint region were identified in DNA from three NF1 individuals (Fig. 8.7).

The largest deletion was approximately 190 kb, and it was entirely contained within the 290–390 kb NotI fragment that encompasses the two translocation breakpoints. Although its ends were not precisely mapped, this deletion removed all three previously described genes between the transloca-tion breakpoints. The second deletion was 40 kb; it encompassed the t(17;22) translocation breakpoint, EVI2A, and the 5' exon of EVI2B. The smallest of the three deletions removed an 11 kb segment of genomic DNA without disrupting any of the known genes. One end of this deletion mapped very close to the t(17;22) translocation breakpoint. Thus, at this point two translocations and three deletions at this locus were known to occur in NF1 individuals, yet none of the rearrangements specifically interrupted a single known gene; therefore, one could not identify which, if any, of the three genes was associated with NF1.

This observation brings up a critical point with respect to isolating disease genes by the mapping approach. The complexity of genetic material con-tained within the human genome is only now being fully appreciated. It appears that the estimated 100 000 genes are not spread out in regular intervals throughout the genome but are sometimes clustered. Reliance on genetic linkage analysis and chromosomal rearrangements to identify a disease locus may in fact uncover a number of genes, all of which are potential candidates for bearing disease-causing mutant alleles. The thoroughness of mutational analysis is *paramount* in the identification of disease genes; the most convincing evidence must come from the identification of single base pair substitutions in the constitutional DNA of affected individuals because,

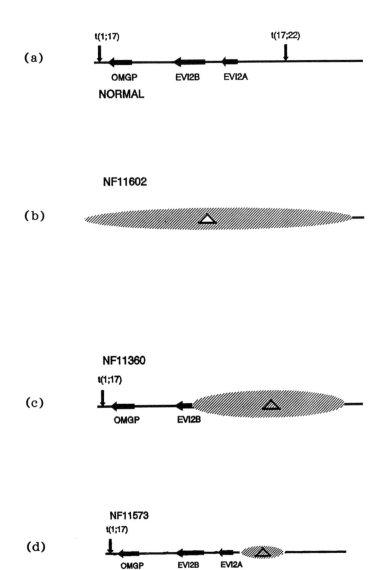

Fig. 8.7. Representation of three deletions of the translocation breakpoint region in alleles of NF1 patients. (a) Normal. (b) NF11602 DNA demonstrating a 190 kb deletion involving all three genes and both translocation breakpoints. (c) NF11360 DNA demonstrating a 40 kb deletion involving *EVI2A* and the 5′ end of *EVI2B*. (d) NF11573 DNA demonstrating an 11 kb deletion involving only the t(17;22) breakpoint

as we have seen with NF1, large deletions may involve more than one gene, and because translocations may affect expression of genes some distance away from the breakpoint. Therefore, chromosomal rearrangements should be used only to identify the disease locus; candidate genes at that locus need to be examined at the level of DNA sequence in order to identify alleles with point mutations. One conclusion to be drawn is that if one does not show convincing evidence of point mutations in candidate genes isolated from a physically defined locus, then a more thorough search for other candidate genes should be undertaken. This constraint certainly applied in the case of the complex *NF1* locus.

IDENTIFICATION OF THE NF1 GENE

During attempts to identify genes at the *NF1* locus, DNA sequences from around the translocation breakpoints were examined in detail for conservation in different animal species. A 3.8 kb *Eco*RI restriction fragment that spanned the t(17;22) breakpoint was subcloned from one of the cosmids (Fig. 8.5), and its DNA sequence was determined. Oligonucleotide primers were synthesized from the sequence, with the aim of pinpointing the breakpoint by PCR analysis. These experiments, which were done using the rodent human hybrid cell line that contained the translocation chromosome, revealed a striking conservation of sequence within the 3.8 kb fragment between human and rodent DNA. Southern blot analysis using the human 3.8 kb fragment as a probe showed a conserved 3.35 kb *Eco*RI band in rodent DNA. This conservation of sequence signalled the possibility of another candidate gene at the *NF1* locus. After several abortive attempts to identify human cDNAs directly using the cosmids as probes, a cDNA clone was identified from a mouse macrophage library. This cDNA, which was shown to identify homologous human sequences in the 3.8 kb *Eco*RI fragment, was used to isolate numerous human cDNA clones from a fetal brain library. As shown in Fig. 8.8, the cDNA sequences mapped to both sides of the t(17;22) breakpoint and were partially deleted in the 11 kb *NF1* deletion described above. This gene was designated TBR (translocation breakpoint region), and because it was the only one of the four genes so far identified that was interrupted by a translocation breakpoint, it became the most likely candidate for *NF1*. But because chromosomal rearrangements can alter gene expression without directly interrupting an exon, an intense search for point mutations in the TBR gene was undertaken, in hopes of identifying mutations in NF1 patients that would not alter expression of other genes at the *NF1* locus.

Since a significant portion of the genomic sequence corresponding to the TBR cDNA was already known, the boundaries of several exons within the genomic DNA were readily defined by comparison of genomic DNA sequence to the cDNA sequence (Cawthon *et al.*, 1990b). A total of nine exons comprising 1991 bp of cDNA, spanning 19.2 kb of genomic sequence, were

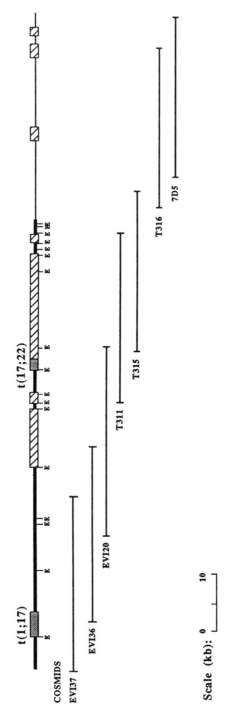

Fig. 8.8. An *EcoRI* map of the NF1 region. The cosmid contig used to construct the map is shown below. The translocation breakpoint regions are shaded and the *EcoRI* fragments hybridizing to the TBR cDNA are represented as hatched areas (Viskochil *et al*, 1990; © Cell Press)

unequivocally mapped at the level of DNA sequence. The end of the open reading frame was identified well downstream of the last base pair match between the genomic and cDNA sequences, and the open reading frame remained open well beyond the most 5' exon mapping to the cosmid contig that represented the genomic DNA spanning both translocation breakpoints (see Fig. 8.9). Several single base pair mutations expected to render the predicted peptide non-functional were found in the coding regions of genomic DNA derived from NF1 patients (Cawthon *et al.*, 1990b). Perhaps the two most important were base substitutions located in exon 4 (arbitrary nomenclature)—one leading to a premature stop and the other substituting a proline for a leucine in the amino acid chain. These findings, together with the previous evidence that the new transcript crossed the t(17;22) breakpoint and the inference that the transcript crossed the t(17;17) breakpoint as well, led to the conclusion that the *NF1* gene had been identified (Cawthon *et al.*, 1990b; Viskochil *et al.*, 1990).

Independent support for this interpretation was provided by the concomitant description of two other cDNA clones, one of which hybridized to the 3.8 kb *Eco*RI fragment containing the t(17;22) breakpoint (Wallace *et al.*, 1990). These clones had been ascertained with a 'jump' clone, derived from an end of one of the *EVI2A* cosmids, EVI20 (O'Connell *et al.*, 1990). In addition, a 500 bp insertion present in an NF1 individual and not in his parents was shown to map to the same *Eco*RI fragment. Although it was ultimately shown that the transcript defined by these two cDNA clones encompassed neither the t(17;22) breakpoint nor the site of the 500 bp insertion, the sequence data indicated that the transcript was identical to the one reported by Cawthon

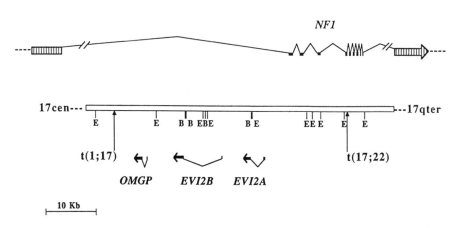

Fig. 8.9. Representation of the *NF1* gene on the restriction enzyme map of the translocation breakpoint region. *NF1* spans both translocation breakpoints and three expressed genes lie in a single intron of the *NF1* gene. Transcriptional orientation is depicted by arrowheads (5'→3'). (Xu *et al*, 1990; © Cell Press)

et al. (1990b). Thus, the 500 bp insertion mutation must lie within the *NF1* gene. Subsequent experiments have shown that the insertion is an *AluI* repeat located in an intron of the *NF1* gene; its functional effect is to alter the splicing pattern in such a way that an exon is dropped from the processed transcripts.

Although the expression level of the NF1 transcript is quite low, it appears to be ubiquitous in both human and mouse tissues. The initial reports describing the *NF1* gene were based on approximately one-half to two-thirds of the full-length cDNA sequence, and the deduced amino acid sequence failed to identify homologies with other previously described proteins in protein sequence databases. Subsequent work over the ensuing years further defined both the physical structure of the *NF1* gene and the functional structure of the *NF1* gene product.

PHYSICAL STRUCTURE OF THE *NF1* GENE

It is now apparent that *NF1* is relatively large, with a transcript estimated at 11–12 kb that spans 335 kb of genomic DNA (Li *et al.*, 1995); its open reading frame specifies a peptide containing 2818 amino acids (Marchuk *et al.*, 1991). There are 59 exons inclusive of two alternatively spliced exons, 23a and 48a. The 5' portion of *NF1* maps to the CpG island of hypomethylation corresponding to the centromeric end of the 350 kb *Not*I fragment shown in Fig. 8.1, and the promoter region and 5'-untranslated region have been sequenced and partially characterized (Hajra *et al.*, 1994). The 3'-untranslated region extends 3.5 kb telomeric of the stop codon, placing the end of the gene approximately 15 kb upstream of the telomeric *Not*I site (Li *et al.*, 1995). The t(1;17) translocation maps to intron 27b, the intron harbouring the embedded genes, and the t(17:22) translocation maps to intron 31. With more complete characterization of the *NF1* locus, the exons lying downstream of the t(1;17) translocation breakpoint named exons 1–9 (Cawthon *et al.*, 1990) have been renumbered beginning with exon 28 (Li *et al.*, 1995). Genomic DNA represented by the contig shown in Fig. 8.8, which encompasses *NF1* exon 28 through the end of the gene, has been sequenced (Weiss *et al.*, 1991).

Intron 27b of the *NF1* gene encompasses all three genes previously mapped between the translocation breakpoints *EVI2A*, *EVI2B* and *OMGP*, but *NF1* is transcribed from the opposite strand with respect to the three interdigitated genes. This phenomenon has no precedent in human genetic literature other than one report of a multiple-copy gene of unknown function that is embedded within an intron of the Factor VIII gene. Interdigitated genes have been reported in *Drosophila*; however, control mechanisms for transcriptional regulation have not been demonstrated. The normal regulation of gene expression for the single-copy, interdigitated genes at the *NF1* locus remains to be investigated. Moreover, how the expression of these three genes might

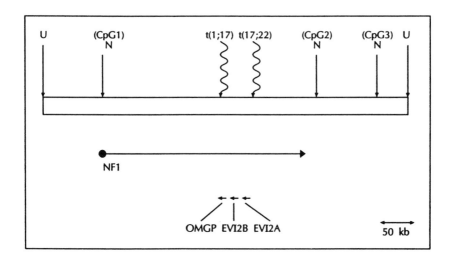

Fig. 8.10. Pulse-field map of the *NF1* locus. *NF1*, *OMGP*, *EVI2B* and *EVI2A* are shown below the genomic fragment. U, *NRUI*; N, *NotI*; CpG, island of hypomethylation. The entire *NF1* gene is represented on this map; its 5′-end is shown as a circle lying just centromeric to CpG island 1 and its 3′-end is contained on the same *NotI* fragment

be affected by mutations in *NF1*, and possibly account for variability in expression of the disease, is unresolved.

Mutational analysis of the *NF1* locus in NF1 patients has been carried out by a number of research groups. Initially, mutations were identified in fewer than 10% of these patients. The known mutations include deletions (Viskochil *et al.*, 1990; Upadhyaya *et al.*, 1990), insertions (Wallace *et al.*, 1990), point mutations (Cawthon *et al.*, 1990b; M. Wallace, personal communication) and translocations. An exhaustive search for mutations in a subset of 40 NF1 patients by SSCP PCR, covering 95% of the coding region of *NF1*, has identified one additional SSCP variant that is a probable mutation (R. Cawthon, personal communication). This unforeseen difficulty in identifying mutations in the *NF1* gene is hard to interpret. The possibility that a yet unspecified and untested common site for mutations exists, as in the cystic fibrosis gene, has not been excluded. However, NF1 is an autosomal-dominant condition with a high spontaneous mutation rate; unless the gene structure were to predispose mutations to occur in a specified region, or 'hot spot', one would expect mutations throughout the entire length of the gene rather than in one region. Furthermore, Richard Cawthon has identified an exon-based polymorphism, and he has demonstrated that both alleles are transcribed from lymphoblastoid cell lines in more than 90% of the NF1 patients who are heterozygous at this locus. This means there are no 'hot spots' in the promoter region of the *NF1* gene that would lead to decreased expression of one allele. Increasingly sensitive screening methodology may need to be developed to identify all the mutations causing this condition. On the other hand, learning something about the function of the NF1 protein may identify a more appropriate screening process for *NF1* mutations.

CHARACTERISTICS OF THE NF1 GENE PRODUCT

An initial search of protein sequence databases for similarities with the peptide predicted from the first 4000 bp of the *NF1* transcript sequence revealed no strong similarities to known proteins. An additional 3000 bp of sequence lying 5′ to this initial segment, however, revealed significant homology in amino acid sequence between the predicted *NF1* product, neurofibromin, and the catalytic domains of the mammalian *ras*-GTPase-activating protein (GAP) and its yeast counterparts, *IRA1* and *IRA2* (Xu *et al.*, 1990a; Buchberg *et al.*, 1990). Futhermore, examination of sequences outside the catalytic domains of the *IRA1* and *IRA2* genes indicated that the similarities between the yeast products and neurofibromin extended for some 300 amino acids in the N-terminal direction, and more than 800 amino acids in the C-terminal direction (see Fig. 8.11). The implications of these similarities are profound, as this family of genes is known to be involved in the mechanisms that control cell growth and differentiation through their interaction with the *ras* gene family.

A number of ras-like proteins are known; within the human p21*ras* subgroup there are three genes lying on separate chromosomes: N-*ras*, H-*ras*

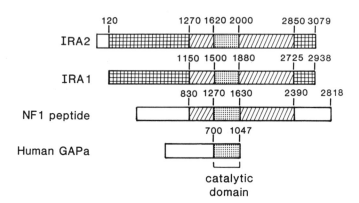

Fig. 8.11. Comparison of the homology between IRA1, IRA2, NF1 protein and GAP. The amino acid sequences deduced from the entire open reading frames are shown as rectangular boxes. Shared homologies as determined by pairwise comparisons are depicted as checkered, hatched and stippled regions. The numbers above the boxes demark the approximate amino acid position for each of the homology boundaries. The 'catalytic domain' of GAP is shown as a bracket

and K-*ras*. Each gene encodes a 21 000 Da protein that hydrolyses GTP to GDP, and they are associated with the inner surface of the cell membrane through their carboxyl termini. The role p21*ras* plays in cellular processes is not completely understood; however, it has been shown to function in specific signal transduction pathways. The oncogenic nature of *ras* mutations has been studied for many years, yet investigators continue to hypothesize about the effectors that transduce 'activated' *ras* signals to pathways leading to cellular transformation. 'Activated' *ras* is associated with human malignancies and it is generally accepted that p21*ras* is active in its GTP-bound state and inactive in its GDP-bound state. GAP substantially accelerates the hydrolysis of *ras*-bound GTP to GDP, thereby converting the *ras* protein from the 'active' GTP-bound form to the 'inactive' GDP-bound form. In view of the amino acid homology between NF1 peptide and the catalytic domain of GAP, it was intriguing to speculate that *NF1* mutations may disrupt the normal interaction of neurofibromin with the *ras* gene product and cause abnormal cell growth. To substantiate this hypothesis rigorous proof of functional homology, through biochemical and genetic experiments, was needed to establish the relationship of the *NF1* gene product to p21*ras* signal transduction pathways.

As predicted, the catalytic domain of neurofibromin is indeed capable of interacting with both mammalian *ras* and yeast *RAS* gene products (Ballester *et al.*, 1990; Martin *et al.*, 1990; Xu *et al.*, 1990b). As shown by brackets in Fig. 8.11, an *NF1* peptide segment comprised of 474 amino acids that show sequence similarity with the *GAP* catalytic domain (GAP-related domain, or GRD) was expressed in a baculovirus–insect ovary protein expression system as NF1-GRD (NF1 gap-related domain) and it stimulated the intrinsic GTPase of p21*ras in vitro* (Martin *et al.*, 1990). NF1-GRD demonstrated a distinct pattern of activity when compared with GAP; it had 30-fold less specific activity *in vitro* but it bound to p21*ras* with 20-fold higher affinity at high p21*ras* concentrations (see Fig. 8.12). The functional homologies were substantiated by further studies demonstrating complementation of *IRA1* and *IRA2* mutants in yeast (Ballester *et al.*, 1990; Martin *et al.*, 1990; Xu *et al.*, 1990b). The identification of two proteins capable of activating the intrinsic GTPase of p21*ras* necessitated the introduction of nomenclature to describe GAP activity from either p120-GAP (120 kDa GTPase *activating protein*) or NF1-GAP. In the course of evaluating p120-GAP and NF1-GRD activities *in vitro*, differences in p21*ras* GTPase activation were noted in response to lipid inhibition, and these studies were extended to reveal that NF1-GAP activity can be distinguished from p120-GAP activity in mammalian cell extracts. Bollag and McCormick (1991) speculated that p120-GAP and NF1-GAP send distinct signals after interaction with p21*ras* and that each may be involved in a separate biochemical signal transduction pathway.

The functional homologies between neurofibromin, p120-GAP and the yeast IRA proteins suggest two general possibilities for the molecular

(a)

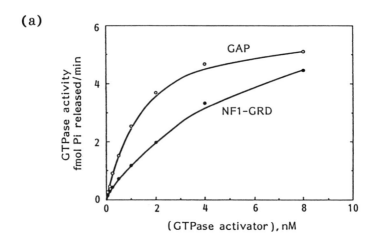

(b)

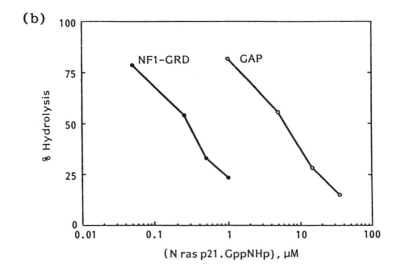

Fig. 8.12. Functional comparison between NF1-GRD and full-length GAP (Martin *et al.*, 1990). Proteins have been purified from the baculovirus/sf9 cell expression system. (a) Stimulation of p21*ras* GTPase at increasing concentrations of GTPase activator and 6 nM GTP-bound p21*ras*. GTPase activity was determined by hydrolysis of the terminal phosphate. (b) Competitive inhibition by ras-p21-GppNHp (a non-hydrolysable GTP analogue) on GTPase activation by NF1-GRD and p120GAP. NF1-GRD has a higher affinity for p21*ras* than full-length GAP yet it has a lower specific activity for activation of ras-GTP hydrolysis *in vitro* (© Cell Press, reproduced by permission)

pathophysiology of NF1. As seen in Fig. 8.13(a), the normally functioning *NF1* genes might downregulate a *ras*-mediated growth signal; in the other (Fig. 8.13b), the product of the normal *NF1* gene would serve as an effector of a *ras*-mediated differentiation signal. In either case, loss of the NF1 product could result in abnormal cell growth. Either hypothesized function, by analogy with the retinoblastoma gene, could be consistent with enhancement of the mutant phenotype through loss of the normal *NF1* allele. Under the hypothesis that the major role of the *NF1* gene is to provide downregulation of a growth signal, the presence of the non-mutant allele might provide nearly normal levels of gene activity and, therefore, near-normal regulation of p21*ras* activity. Somatic loss or mutation of the non-mutant allele might be necessary for the clinical expression of severe medical manifestations such as

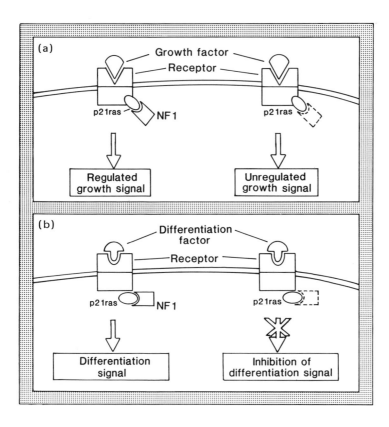

Fig. 8.13. Two models for the pathophysiology of a defect in *NF1*. (a) NF1 protein as a negative regulator of the growth stimulation pathway whereby loss of functional NF1 protein would permit unregulated growth. (b) NF1 protein as an effector in the differentiation pathway whereby loss of functional NF1 protein would inhibit cell differentiation

malignancies. Under the hypothesis that the major role of *NF1* is to propagate a differentiation signal, the presence of the normal allele might provide adequate gene activity to propagate the signal. If this model were correct, emergence of the mutant phenotype in a cell might also require loss of the normal allele.

Perhaps the most accessible way to study the tumour suppressor nature of the NF1 protein is by evaluating loss of heterozygosity (LOH) for certain polymorphic loci in tumours found in NF1 patients. In contrast to the consistent LOH seen in retinoblastomas at the Rb locus, the evidence supporting chromosomal loss in NF1 tumours has been equivocal. Initial studies examining neurofibromas in NF1 patients for loss of chromosomal material failed to demonstrate loss of heterozygosity near band 17q11.2 (Skuse *et al.*, 1989). Given that neurofibromas are a mixture of cell types, possibly originating from different embryological sources, failure to demonstrate LOH at the *NF1* locus could be explained if neurofibromas were polyclonal. However, subsequent studies have demonstrated both the monoclonal nature of neurofibromas and absence of LOH with respect to informative markers *near* the *NF1* locus. Still, one cannot reject the tumour suppressor model for neurofibroma development on this evidence alone. The basic tenet of such a model is inactivation of a remaining normal allele; this can occur through chromosomal loss or by point mutation in the tumour suppressor gene. Therefore, until the normal allele in monoclonal neurofibromas is shown to be functional, one cannot exclude the hypothesis that NF1 protein is a tumour suppressor. In contrast to neurofibromas, malignant tumours associated with NF1 have demonstrated variable losses of chromosome 17 material. In neurofibrosarcomas, complete loss of chromosome 17 or partial loss of 17p has been demonstrated (Menon *et al.*, 1990), whereas specific loss of 17q material near the *NF1* locus has been noted in a small number of neurofibrosarcomas and in an astrocytoma (Skuse *et al.*, 1989). The latter study, however, did not evaluate LOH at the p53 locus on 17p in malignant tumours. Another study demonstrated specific loss of 17q in phaeochromocytomas from *some* NF1 patients, yet failed to document loss of 17q chromosomal material in phaeochromocytomas obtained from any of the non-NF1 individuals (Xu *et al.*, 1992). These studies are only suggestive of NF1 protein acting as a tumour suppressor. Now that intragenic probes and DNA sequence from *NF1* are available, one can test the tumour suppressor hypothesis by examining the *NF1* locus in detail for inactivation of the normal allele in NF1 patient material. Furthermore, one can now evaluate NF1-GAP function in cell lysates from NF1 tumours, and soon it should be possible to test for protein expression with anti-NF1 antibody.

Inactivation of the normal allele in NF1 patients is a feasible model for tumour formation, and the interaction of NF1 GAP-related domain with p21*ras* provides a plausible mechanism for abnormal cell growth. However, CLS, Lisch nodules, learning disabilities and skeletal dysplasias do not readily provide simple explanations for the pathophysiology of NF1 with respect to the p21*ras* pathway. The fact that few *NF1* mutations have been observed in the GAP-related domain suggests that this region may play only a small role in the classical NF1 phenotype and perhaps, except for presentations involving malignancies, the domains outside of the GRD strongly influence the clinical expression of this condition. In support of this interpretation, nearly 99% homology exists between rodent and human predicted amino acid sequence over the entire length of the *NF1* peptide (Buchberg *et al.*, 1990; Bernards *et al.*, 1993). It is plausible that as yet unidentified, tissue-specific proteins associate with different domains in neurofibromin and modify the cellular effects of *NF1*-p21*ras* interaction. If true, such observations would help explain the high degree of clinical variability in NF1.

The protein products encoded by the alternate splice forms of the *NF1* transcript may also play a significant role in the NF1 phenotype. Analysis of the developmental and tissue-specific expression of the various transcripts and their respective peptide functions may delineate alternative *NF1* signal transduction processes underlying the variability of the NF1 phenotype. For example, one of the alternate splice forms involves an exon in the *NF1*-GRD. During screening of lymphoblastoid RNA-directed PCR products for mutations in *NF1*, a slower-migrating product was identified in agarose gels (R. Cawthon, personal communication) in addition to the product of the size predicted on the basis of cDNA sequence. Sequencing of this larger fragment revealed an in-frame, 63 bp exon that encoded a peptide enriched in lysine residues (see Fig. 8.14). The significance of this alternate splice form is being evaluated but one can speculate that, as in other nervous tissue, differential splicing is tissue-specific and may be restricted in developmental expression during embryogenesis. The implications of this hypothesis with respect to the variable clinical presentation of NF1 are obvious but untested. Mutations in *NF1* leading to abnormal splicing may not be detected unless screening involves transcripts from the appropriate tissue at the appropriate stage in organ development.

CLINICAL APPLICATIONS OF THE ISOLATED NF1 GENE

The clinical applications resulting from the identification and characterization of the *NF1* gene have yet to come to fruition. This is in part due to a number of investigational hurdles, which include the current lack of identified mutations in the gene in most of the NF1 patients; the fact that no defined cell type presents a distinct phenotype associated with NF1; the slowness of progress toward the development of a good antibody against NF1 protein; and an inability to predict which NF1 patients will develop more severe medical problems. Nevertheless, prenatal diagnosis is now available for families where genetic linkage can be performed, and individuals at risk be tested

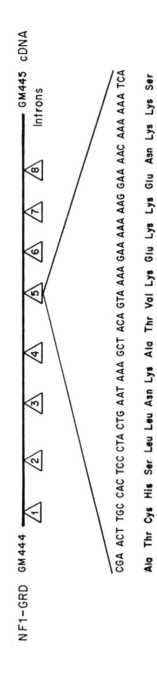

Fig. 8.14. Alternate splice form in the GAP-related domain of NF1. The exon sequence of the GAP-related domain (as defined by the construct using primers GM 444 and GM 445k (Martin *et al.*, 1990)) is depicted as a bold line and intron sequence is represented as numbered triangles. There are eight introns in the GAP-related domain of *NF1*. The fifth intron contains an exon that is expressed as an alternate splice form of the NF1 transcript. The exon is 63 bases and encodes an amino acid fragment rich in lysine residues, and it corresponds to *NF1* exon 23a (Li *et al.*, 1995)

if other family members have NF1. The future holds promise that once mutations are identified, a genotype/phenotype correlation study can be carried
out in the hope of providing families with better prediction and clinical guid-
ance. A number of clinical conditions with NF1 features have been termed
'NF overlap syndromes'. The development of a full mutational screen of *NF1*
would enable clinicians to distinguish variable expression of the NF1 phenotype from other variant forms of neurofibromatosis. Many of the unusual
clinical presentations of NF1 could be evaluated in this way and placed in one
of three general categories: allelic forms of NF1, mutations involving more
than one gene at the *NF1* locus (such as *EVI2A*, *EVI2B* or *OMGP*), or mutations affecting genes encoding other as yet unknown proteins that may share
a biochemical pathway with neurofibromin. For many NF-related conditions
the identification of *NF1* gene mutations would be important for genetic
counselling because often the NF overlap syndromes are not inherited whereas NF1 carries a risk of 50% for each offspring born to an affected parent.

The association of NF1 with malignancies is a compelling incentive for the
development of therapeutic regimens to control benign neurofibromas and
other tumours that may predispose to cancer progression. To date there are
no medications available for effective treatment of the clinical features of NF1;
most medical complications are dealt with as they arise, with variable
outcomes. The isolation of the *NF1* gene and the demonstration of the
encoded peptide's association with p21*ras* have identified a heretofore
unknown biochemical pathway involving lipid regulation of p21*ras* GTP
hydrolysis. Characterization of this pathway may provide opportunities for
the development of specific therapeutic intervention in NF1.

CONCLUDING REMARKS

The discovery of the *NF1* gene by the mapping approach, and characterization of the NF1 gene product, offer the promise of new means for the
diagnosis and study of this disease. New research approaches may ultimately
lead to an effective means of medical intervention. As is often the case in the
study of rare human genetic disorders, however, the final significance may be
even broader because the gene has now been specifically placed within the
category of proteins that act in cells to control growth and differentiation. The
NF1 protein is, therefore, of great interest to investigators worldwide who are
seeking to understand normal cellular mechanisms as well as the altered
signal transduction pathways that lead to the formation of tumours.

ACKNOWLEDGEMENTS

I thank R. Foltz for editing the manuscript, J. Carey MD for his helpful
suggestions, and P. O'Connell and R. Cawthon for Figure development.

REFERENCES

Ballester R, Marchuk D, Boguski M et al. (1990) The NF1 locus encodes a protein functionally related to mammalian GAP and yeast IRA proteins. Cell, 63, 851–859.

Barker D, Wright E, Nguyen K et al. (1987) Gene for von Recklinghausen neurofibromatosis is in the pericentromeric region of chromosome 17. Science, 236, 1100–1102.

Bernards A, Snijders A, Hannigan G et al. (1993) Mouse neurofibromatosis type 1 cDNA sequence reveals high degree of conservation of both coding and noncoding mRNA segments. Hum. Mol. Genet., 2, 645–650.

Bollag G, and McCormick F (1991) Differential regulation of rasGAP and neurofibromatosis gene product activities. Nature, 351, 576–579.

Buchberg A, Cleveland L, Jenkins N and Copeland N (1990) Sequence homology shared by neurofibromatosis type-1 gene and IRA-1 and IRA-2 negative regulators of the RAS cyclic AMP pathway. Nature, 347, 291–294.

Cawthon R, O'Connell P, Buchberg A et al. (1990a) Identification and characterization of transcripts from the neurofibromatosis 1 region: the sequence and genomic structure of EVI2 and mapping of other transcripts. Genomics, 7, 555–565.

Cawthon R, Weiss R, Xu G et al. (1990b) A major segment of the neurofibromatosis type 1 gene: cDNA sequence, genomic structure, and point mutations. Cell, 62, 193–201.

Cawthon R, Anderson L, Buchberg A et al. (1991) cDNA sequence and genomic structure of EVI2B, a gene lying within an intron of the neurofibromatosis type 1 gene. Genomics, 9, 446–460.

Fountain J, Wallace M, Bruce M et al. (1989) Physical mapping of a translocation breakpoint in neurofibromatosis. Science, 244, 1085–1087.

Hajra A, Martin-Gallardo A, Tarle S et al. (1994) DNA sequences in the promoter region of the NF1 gene are highly conserved between human and mouse. Genomics, 21, 649–652.

Lefkowitz I, Obringer A and Meadows A (1990) Neurofibromatosis and cancer: incidence and management. In Rubenstein A and Konf B (eds), Neurofibromatosis: A Handbook for Patients, Families and Health-care Professionals (pp. 99–110). New York: Theime.

Li Y, O'Connell P, Huntsman Breidenbach H et al. (1995) Genomic organization of the neurofibromatosis 1 gene (NF1). Genomics, 25, 9–18.

Marchuk D, Saulino A, Tavakkol R et al. (1991) cDNA Cloning of the type 1 neurofibromatosis gene: complete sequence of the NF1 gene product. Genomics, 11, 931–940.

Martin G, Viskochil D, Bollag G et al. (1990) The GAP-related domain of the neurofibromatosis type 1 gene product interacts with ras p21. Cell, 63, 843–849.

Menon A, Anderson K, Riccardi V et al. (1990) Chromosome 17p deletions and p53 gene mutations associated with the formation of malignant neurofibrosarcomas in von Recklinghausen neurofibromatosis. Proc. Natl Acad. Sci. USA, 87, 5435–5439.

O'Connell P, Leach R, Ledbetter D et al. (1989a) Fine structure mapping studies of the chromosomal region harboring the genetic defect in neurofibromatosis type 1. Am. J. Hum. Genet., 44, 51–57.

O'Connell P, Leach R, Cawthon R et al. (1989b) Two NF1 translocations map within a 600-kilobase segment of 17q11.2. Science, 244, 1087–1088.

O'Connell P, Viskochil D, Buchberg A et al. (1990) The human homologue of murine evi-2 lies between two translocation breakpoints associated with von Recklinghausen neurofibromatosis. Genomics, 7, 547–554.

Seizinger B, Rouleau G, Ozelius L et al. (1987) Genetic linkage of von Reckinghausen

neurofibromatosis to the nerve growth factor receptor gene. *Cell*, **49**, 589–594.
Skuse G, Kosciolek B and Rowley P (1989) Molecular genetic analysis of tumors in von Reckinghausen neurofibromatosis: loss of heterozygosity for chromosome 17. *Genes Chromosomes Cancer*, **1**, 36–41.
Stumpf D, Alksne J, Annegers J *et al.* (1988) Neurofibromatosis. *Arch. Neurol.*, **45**, 575–578.
Upadhyaya M, Cherryson A, Broadhead W *et al.* (1990) A 90kb DNA deletion associated with neurofibromatosis type 1. *J. Med. Genet.*, **27**, 738–741.
Viskochil D, Buchberg A, Xu G *et al.* (1990) Deletions and a translocation interrupt a cloned gene at the neurofibromatosis type 1 locus. *Cell*, **62**, 187–192.
Viskochil D, Cawthon R, O'Connell P *et al.* (1991) The gene encoding the oligoden-drocyte-myelin glycoprotein is embedded within the neurofibromatosis type 1 gene. *Mol. Cell. Biol.*, **11**, 906–912.
Wallace M, Marchuk D, Anderson L *et al.* (1990) Type 1 neurofibromatosis gene: identification of a large transcript disrupted in three NF1 patients. *Science*, **249**, 181–186.
Weiss R, Dunn D, DiSera L *et al.* (1992) The human neurofibromatosis type 1 locus: genomic sequence of the 3' region, 1–100849, *Genbank* Accession Number L05367.
Xu G, O'Connell P, Viskochil D *et al.* (1990a) The neurofibromatosis type 1 gene encodes a protein related to GAP. *Cell*, **62**, 599–608.
Xu G, Lin B, Tanaka K *et al.* (1990b) The catalytic domain of the neurofibromatosis type 1 gene product stimulates *ras* GTPase and complements *ira* mutants of S. *cerevisiae*. *Cell*, **63**, 835–841.
Xu W, Mulligan L., Ponder M *et al.* (1992) Loss of alleles in pheochromocytomas from patients with type 1 neurofibromatosis. *Genes Chromosomes Cancer*, **4**, 337–342.

FURTHER READING

Bos J (1988) The *ras* gene family and human carcinogenesis. *Mutat. Res.*, **195**, 255–271.
Botstein D, White R, Skolnick M and Davis R (1980) Construction of a genetic linkage map in man using restriction fragment length polymorphisms. *Am. J. Hum. Genet.*, **32**, 314–331.
Bourne H, Sander D and McCormick F (1991) The GTPase superfamily: conserved structure and molecular mechanism. *Nature*, **349**, 117–127.
Chen C, Malone T, Beckendorf S and Davis R (1987) At least two genes reside within a large intron of the *dunce* gene of *Drosophila*. *Nature*, **329**, 721–724.
Crowe F, Schull W and Neel J (1956) *A Clinical, Pathological, and Genetic Study of Multiple Neurofibromatosis*. Springfield, IL: Charles C. Thomas.
Declue J, Cohen B and Lowy D (1991) Identification and characterization of the neurofibromatosis type 1 protein product. *Proc. Natl Acad. Sci. USA*, **88**, 9914–9918.
Declue J, Stone J, Blanchard R *et al.* (1991) A *ras* effector domain mutant which is temperature sensitive for cellular transformation: interactions with GTPase-activating protein and NF-1. *Mol. Cell. Biol.*, **11**, 3132–3138.
Goldgar D, Green P, Parry D and Mulvihill J (1989) Multipoint linkage analysis in neurofibromatosis type 1: an international collaboration. *Am. J. Hum. Genet.*, **44**, 6–12.
Golubic M, Tanaka K, Dobrowolski S *et al.* (1991) The GTPase stimulatory activities of the neurofibromatosis type 1 and the yeast IRA2 proteins are inhibited by arachidonic acid. *EMBO J.*, **10**, 2897–2903.
Gutmann D, Wood D and Collins F (1991) Identification of the neurofibromatosis type 1 gene product. *Proc. Natl Acad. Sci. USA*, **88**, 9658–9662.

Hall A (1990) *ras* and GAP—who's controlling whom? *Cell*, **61**, 921–923.

Han J-W, McCormick F and Macara I (1991) Regulation of Ras-GAP and the neurofibromatosis-1 gene product by eicosanoids. *Science*, **252**, 576–579.

Henikoff S and Eghtedarzadeh M (1987) Conserved arrangement of nested genes at the *Drosophila Gart* locus. *Genetics*, **117**, 711–725.

Henikoff S, Keene M, Fechtel K and Fristrom J (1986) Gene within a gene: nested *Drosophila* genes encode unrelated proteins on opposite DNA strands. *Cell*, **44**, 33–42.

Leach R, Thayer M, Schafer A and Fournier REK (1989) Physical mapping of human chromosome 17 using fragment-containing microcell hybrids. *Genomics*, **5**, 167–176.

Mikol D, Alexakos M, Bayley C et al. (1991) Structure and chromosomal localization of the gene for the oligodendrocyte-myelin glycoprotein. *J. Cell Biol.*, **111**, 2673–2679.

Riccardi V and Eichner J (1986) *Neurofibromatosis: Phenotype, Natural History, and Pathogenesis.* Baltimore: Johns Hopkins University Press.

Santos E and Nebreda A (1989) Structural and functional properties of *ras* proteins. *FASEB J.*, **3**, 2151–2163.

Yagel M, Parruti G, Xu W, Ponder B and Solomon E (1990) Genetic and physical map of the von Reckinghausen neurofibromatosis (NF1) region on chromosome 17. *Proc. Natl Acad. Sci. USA*, **87**, 7255–7259.

9 The Genetics of Psychiatric Disorders

PHILIP J. ASHERSON and MICHAEL J. OWEN

The subject of this chapter is the application of molecular genetics to the study of psychiatric disorders. We have chosen to focus upon the dementias and the so-called 'functional' psychoses, schizophrenia and manic-depression, because these are the main causes of severe mental illness. Unlike many of the diseases discussed in this volume they are very common, each affecting over 1% of the population. As we shall see, genetic factors play an important role in their causation, but like other common disorders they do not appear to be inherited in a simple Mendelian fashion. This greatly complicates attempts to apply molecular genetic techniques. In addition, the psychiatric geneticist faces problems of diagnosis and disease definition. These difficulties have meant that progress in this field has not always been easy and overcoming them is one of the greatest challenges facing geneticists.

In the first section of this chapter we shall describe theoretical and methodological issues relating to genetic marker studies in schizophrenia in some detail, since they are also of relevance to the discussions of other diseases that follow.

SCHIZOPHRENIA

THE CLINICAL SYNDROME

Schizophrenia is a common disorder with an incidence of 0.1–0.5 per thousand of the population. The lifetime risk for developing schizophrenia is about 1%. It is the major cause of chronic psychiatric morbidity and is a socially disabling disorder to which until recently more hospital beds were devoted in the UK than to any other single disease. It was first clearly delineated by Kraepelin in 1893, who divided the functional psychoses into dementia praecox (schizophrenia) and manic-depressive insanity (manic-depressive illness)—a division which remains the basis of all modern classifications. Despite the high prevalence rate it has proved one of the most

Molecular Genetics of Human Inherited Disease. Edited by D.J. Shaw
Published 1995 by John Wiley & Sons Ltd

difficult psychiatric syndromes to define, with many widely divergent concepts being held in different countries at different times.

Essentially schizophrenia is characterized by two syndromes that commonly occur in the same individuals. The 'positive syndrome' consists of delusions, hallucinations and disorders of thought and communication, while the 'negative syndrome' is characterized by apathy, lack of drive, slowness and social withdrawal. They differ in their response to medication and outcome in that positive symptoms generally respond to medication and run a relatively acute course, while signs of the negative syndrome do not usually respond to medication and may go on to form long-term deficits which account for many of the social impairments seen in patients with schizophrenia. Attempts have been made to define subtypes based on the relative predominance of these two syndromes and using other validating criteria such as presence of affective (mood) symptoms, course of illness, response to treatment and family history.

More recently neuroimaging with computer tomography (CT) and magnetic resonance imaging (MRI) has provided us with clear evidence of structural brain changes in the form of ventricular enlargement (Fig. 9.1) and localized changes to medial temporal lobe structures. In addition, functional

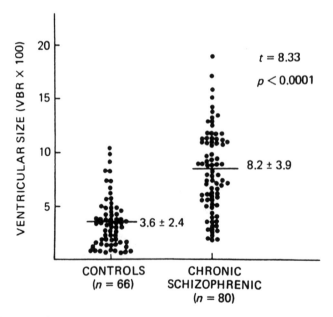

Fig. 9.1. Ventricular size in schizophrenia. VBR, ventricular–brain ratio. From Nasrallah H (1986) Cerebral hemisphere asymmetries and interhemispheric. In Nasrallah HA and Weinberger DR (eds), *The Neurology of Schizophrenia* Amsterdam: Elsevier/North Holland.

brain changes demonstrated by altered regional blood and metabolic changes in the cerebral cortex have been observed in individuals with schizophrenia and there is some evidence that specific functional changes are associated with specific groups of symptoms and cognitive defects.

PROBLEMS OF DIAGNOSIS

DNA marker studies of schizophrenia are complicated by uncertainties in defining the phenotype. This is important in linkage studies since misclassification can lead to spurious results. This type of error is unlikely to produce false evidence of linkage but may reduce the power of an analysis so that linkage may be missed even when it is in fact present.

At present, diagnosis of schizophrenia can only be made by medical examination, relying on the identification of clinical syndromes derived from the patient's account of his mental state and on certain behavioural observations. Through the use of standardized interviews and operationalized diagnostic criteria, we are now able to achieve very good diagnostic reliability. However, the validity of the resulting categorical distinctions are uncertain—a fact reflected by the numerous different definitions of schizophrenia which exist. For example, within families affected with schizophrenia it is common to find related conditions such as schizoaffective disorder, atypical psychosis and schizotypy. These disorders are distinguished from schizophrenia by the precise number and type of psychotic symptoms, the balance of affective (mood) to psychotic symptoms and the longitudinal course of the disorder. What is striking here is the apparent variability in the expression of the disorder, which leads to uncertainty in defining the breadth of the phenotype. An example of this is the difficulty that can arise in distinguishing the two main functional psychoses: manic depression and schizophrenia. The two main syndromes are easy to distinguish, but intermediate forms, such as schizoaffective disorder which has clinical features characteristic of both, commonly occur in families with schizophrenia and manic depression.

One approach to defining the appropriate phenotype is to group together related diagnostic categories and to determine which provides the highest estimates of measures of 'geneticity' such as heritability and monozygote: dizygote twin concordance ratios. In schizophrenia this approach suggests that a broad phenotype exists, including schizophrenia, schizoaffective disorders and schizotypy, but not typical forms of affective disorders.

FAMILY, TWIN AND ADOPTION STUDIES

It has long been recognized that schizophrenia runs in families, and there is strong evidence from twin and adoption studies that genetic factors are important in this disorder. These studies have provided the impetus for progression to the search for aetiologically significant genes using modern molecular and genetic techniques (See Gottesman, 1991).

The first systematic family study was carried out in 1916 and showed that there was a higher rate of dementia praecox (schizophrenia) among the siblings of probands than in the general population. A study of over a thousand schizophrenics carried out in 1938 showed increased rates among siblings and offspring. In more recent studies using operational diagnostic criteria the risk among various groups of relatives remains high (Table 9.1).

Numerous twin studies have been carried out and these show concordance rates for schizophrenia in MZ twins of the order of 50%, compared with around 17% in DZ twins. Consistent with these findings, adoption studies have demonstrated an excess of schizophrenia among the offspring of schizophrenic parents compared with matched controls. This latter finding was supported by a study which took a different approach, in which the rate of schizophrenia was higher among the biological parents of schizophrenic probands who had been adopted, than among their adoptive parents. An interesting finding in this work was that a spectrum of illness was found among the biological parents and this gave rise to the concept of 'schizo-phrenia spectrum disease', which includes both schizophrenia and milder schizophrenia-like conditions. As we have already seen, such variable expressivity can lead to difficulties in deciding which members of families can be designated as 'affected' for the purposes of genetic linkage studies.

Table 9.1. Lifetime expectancy (morbid risk) of schizophrenia in the relatives of schizophrenics

Type of relative	Number at risk (BZ)	Lifetime expectancy (%)	r^a
First-degree			
Parent	8020.0	5.6	0.30
Siblings	9920.7	10.1	0.48
Siblings with one parent schizophrenic	623.5	16.7	0.57
Children	1577.6	12.9	0.50
Children with both parents schizophrenic	134.0	46.3	0.85
Second-degree			
Half siblings	499.5	4.2	0.24
Uncles/aunts	2421.0	2.4	0.14
Nephews/nieces	3965.5	3.0	0.18
Grandchildren	739.5	3.7	0.22
Third-degree			
First cousins	1600.5	2.4	0.14

BZ is the 'risk lives' or sample size.
[a] Correlation in liability assuming general population morbid risk of 1%.
Data combined from many studies by Gottesman II and Shields J (1982) *Schizophrenia: The Epigenetic Puzzle.* Cambridge, UK: Cambridge University Press.

WHAT IS THE MODE OF TRANSMISSION?

Linkage analysis has been successfully applied in identifying chromosomal regions containing genes for diseases such as Huntington's disease, myotonic dystrophy, Friedreich's ataxia and cystic fibrosis. These all have in common a simple Mendelian mode of transmission which is apparent from examining segregation patterns within affected pedigrees and it is relatively easy to estimate the genetic parameters of penetrance and gene frequency required for linkage analysis. In schizophrenia the situation is far less certain since observations of familial risks and familial clustering do not demonstrate simple Mendelian inheritance but irregular or complex patterns of segregation, which could be brought about by a number of different mechanisms (see McGuffin, 1991).

SINGLE-GENE MODELS

In the simplest models single genes are considered to be the sole source of genetic influence resulting in resemblance among relatives. These are termed 'general single locus' (GSL) models. How can they explain the irregular pattern of transmission observed in the functional psychoses? One possible explanation is variable expressivity. An example of this is neurofibromatosis, in which an affected individual may show a single café-au-lait mark (a small discoloured patch on the skin), whereas in others a severe condition with multiple skin tumours and systemic involvement can occur. It has been suggested that schizophrenia might be a single-gene dominant disorder with highly variable expression ranging from a 'core' syndrome through milder schizoid traits (schizophrenic-like features) to a range of minor psychological characteristics among relatives.

Another possible explanation for irregular transmission in a single-gene disorder is non-expression in some individuals who carry the mutant gene. This phenomenon is referred to as reduced penetrance. In fact there is strong evidence of reduced penetrance in schizophrenia which comes from the observation that the lifetime risk for developing schizophrenia among the offspring of unaffected co-twins in discordant monozygotic twin pairs is as high as that in the offspring of the affected co-twin (Table 9.2). In other words, the 'disease gene' has been transmitted but was not expressed in the parent.

While GSL models have been proposed, attempts to test goodness-of-fit with estimates of population frequency of the disease and risks in relatives of various degrees have found no unique solutions. In one study it was found that the best solution for a single-gene model when schizophrenia was treated as a homogeneous disorder was a partially recessive mode of inheritance in which the homozygote for the disease gene has a reduced penetrance.

Table 9.2. The offspring of twins discordant for schizophrenia

	Monozygotic twins		Dizygotic twins	
	Affected	Well	Affected	Well
Morbid risk of schizophrenia in offspring	16.8%	17.4%	17.2%	2.1%

Data from Gottesman and Bertelsen (1989).

MULTIFACTORIAL THRESHOLD MODEL

Under a multifactorial threshold (MFT) model genetic factors are assumed to be oligogenic or polygenic. In other words, several or many genes, each of small effect, combine additively, with or without the effects of environmental factors to influence liability to schizophrenia. Individuals who develop the disorder are considered to lie above a threshold value along a continuum of liability, which is itself normally distributed within the general population (Fig. 9.2). The MFT model of inheritance can account for the observed risks to different classes of relatives which declines exponentially from monozygotic twins, to first, to second and then to third-degree relatives. This is explained by a reduction in genetic risk due to a shift of the liability curve with successive generations, as the number of shared genes reduces from 1 to 1/2 to 1/4 to 1/8 and so on. MFT models therefore fit more easily data on the lifetime risks for schizophrenia among relatives than GSL models.

MIXED MODELS

Mixed models are intermediate between GSL and MFT models and involve the action of both a gene of major effect and polygenes. In other words, the expression of a single major gene is modified by interaction or co-action with a number of other genes, each having only a small effect on their own. Studies of mixed models in schizophrenia have generally failed to find support for a single major gene effect and this has favoured MFT models.

AETIOLOGICAL HETEROGENEITY

So far it has proven impossible to demonstrate clearly which genetic model is most applicable to schizophrenia. It may well be that, as with mental handicap, there is genetic and aetiological heterogeneity so that several different mechanisms are acting on the population to produce overlapping disorders.

Pedigrees can certainly be found which contain a high density of schizophrenic individuals and which have a 'dominant-like' appearance (Fig. 9.3).

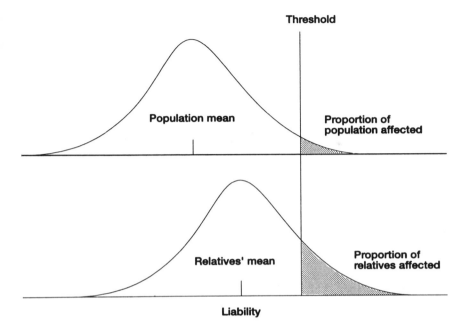

Fig. 9.2. Multifactorial threshold model (MFT). Liability is considered to be normally distributed in the general population and results from the additive action of polygenes and non-inherited factors. First-degree relatives of affected individuals inherit on average half of their genes, thereby increasing their liability to schizophrenia. This is illustrated by a shift in the liability curve to the right, so that more individuals lie above the threshold

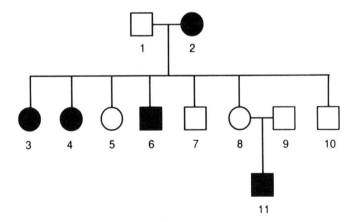

Fig. 9.3. Pedigree of a family with schizophrenia. The mode of transmission *appears* to be Mendelian. For the purposes of linkage analysis person 8 is assumed to be a non-penetrant gene carrier

In these families single genes may be the sole or main source of resemblance between relatives. There may, however, be more than one mutation at a single genetic locus (allelic heterogeneity) and different pedigrees may segregate completely different disease genes (non-allelic heterogeneity). Furthermore, since single genes are unlikely to account for all cases of schizophrenia, aetiological heterogeneity might exist in the form of mixed genetic and environmental effects.

A recent example of heterogeneity in psychiatry is Alzheimer's disease, where mutations of the amyloid precursor protein (APP) gene on chromosome 21 have been shown to segregate with the disease in some familial, early-onset families, while a locus on chromosome 14 has been linked to the disease in other similar families. At the same time these two loci only account for a small proportion of all those with the disease (see below). Other common disorders such as non-insulin-dependent diabetes, coronary artery disease and breast cancer also display this type of aetiological heterogeneity, with a small proportion of cases resulting from a highly penetrant defect in a single gene. However, there is as yet no compelling evidence that subforms of schizophrenia resulting from major gene effects actually exist and research based on this assumption is to some extent still an act of faith (Owen and McGuffin, 1992).

DNA MARKER STUDIES

DNA markers can be used to locate disease genes by linkage and association studies. In linkage studies using the LOD score method, co-segregation of a marker and the disease is examined within families containing several affected members. Departure from independent assortment provides evidence of linkage and enables an estimate of the recombination fraction to be made. In contrast, association studies compare the frequencies of marker alleles between groups of affected individuals and samples of controls without the disease or drawn from the general population. A difference suggests either very tight linkage resulting in linkage disequilibrium between the marker allele and disease mutation, or that the marker allele is itself the mutation conferring susceptibility to disease. Linkage studies are more difficult to carry out since they involve the study of multiply affected pedigrees, which are harder to collect than the unrelated individuals required for association studies. However, linkage is a powerful technique for locating genes of major effect and a single marker may give information over a large genetic distance. In contrast, association studies are able to identify genes of both small and large effect. However, linkage disequilibrium will only exist if the marker and the disease gene are very close (less than 1 cM), and if the mutation rates of both are sufficiently low.

LINKAGE STUDIES

In the LOD score approach to linkage analysis, penetrance values and gene frequency must be specified. Failure to specify these parameters accurately reduces the power to detect linkage and may lead to false-negative findings. LOD score analysis has been enormously successful in locating disease genes in disorders where the mode of transmission follows a simple Mendelian pattern, allowing the accurate estimation of these parameters. However, this is not the case in schizophrenia. This problem may at first sight seem to preclude the successful application of LOD score analysis to schizophrenia, but as we have seen this disorder sometimes occurs in large pedigrees with multiple affected members, which appear to show more regular modes of transmission. Although these loaded families are atypical, and the resulting Mendelian appearance can be misleading, no bias with respect to detection of linkage should be introduced. The assumption implicit in the large-family LOD score approach, adopted by most groups working in this field, is that genes of major effect exist in these families. This assumption can, however, only be proved (or disproved) by systematic screening of the entire genome in an extensive set of families using highly informative genetic markers.

Faced with difficulties of diagnosis and unknown modes of inheritance, how do we proceed with a linkage study? Since the main problem is that linkage may be missed when it in fact exists, a range of diagnostic definitions and genetic parameters can be used in order to hit upon the correct combination as nearly as possible. This exploratory approach does, however, involve the use of multiple genetic and diagnostic models which will inevitably result in an increase in false-positive results. This is reflected in the amount of confidence that can be placed on a given LOD score to indicate the existence of true linkage. It has been the convention to accept an LOD score of 3 as evidence of linkage. However, the increase in false-positive findings when multiple tests are used means that higher odds are required.

It is in fact difficult to estimate the true significance of a LOD score under these circumstances and the best method may be to undertake a complete computer simulation of the study. Of course, a very high LOD score may be beyond dispute, but there are several factors which make this unlikely to occur, at least with the samples available to most centres. The first is that if the true mode of inheritance includes low penetrance, the power to detect linkage is markedly reduced. The second is the possibility of genetic heterogeneity or aetiological heterogeneity in which a defect at a particular genetic locus is responsible for the disease in only a subset of families. Since the overall LOD score in a set of pedigrees comes from adding together the individual scores from each family, the presence of both linked and non-linked families may result in positive and negative scores cancelling each other out. Misspecification of parameters and diagnostic uncertainty is also likely to reduce the power of linkage analysis.

To overcome these problems pedigrees must be of adequate size and informative for linkage. A large number of such pedigrees will be needed to demonstrate linkage in the presence of heterogeneity. Dense maps of highly polymorphic markers which are now available will greatly increase the power of these studies.

Another area of difficulty is the interpretation of negative LOD scores. By convention, a LOD score of -2 at a particular value of θ implies exclusion of the region defined by that θ value. However, as with the interpretation of positive LOD scores, the situation here is more complex. The conventional criteria assume homogeneity, known mode of inheritance and diagnostic certainty. Therefore, in describing a region as 'excluded', it is important to describe under what specific genetic and diagnostic models the exclusion criterion was reached. The effect of locus heterogeneity on exclusion of chromosome 19 is illustrated in Fig. 9.4. Another effect of heterogeneity on exclusion data is described later in this chapter in the description of work on Alzheimer's disease, where false exclusion of the APP gene was produced, prior to the later confirmation of linkage.

MODEL-FREE METHODS

It is possible to avoid some of the problems arising from unknown modes of inheritance by carrying out analyses using so-called 'model-free methods'. In affected sibling-pair methods, the distribution of alleles in affected siblings is compared with that expected under random segregation. On average, siblings would be expected to share two parental alleles 25% of the time, one parental allele 50% of the time and none 25% of the time. An increase in the number of alleles shared between affected siblings would indicate linkage. This test for linkage is appealing in that it avoids the need to make assumptions about the mode of inheritance and the affected sibling pairs can be selected for core symptoms, increasing the certainty that they represent true 'genetic' forms of the disorder. This method also has the considerable advantage that it can identify genes of moderate effect where several genes are acting together. The disadvantages are that very large samples are required, although each family is much easier to collect than the larger families used in most conventional linkage studies. An adaptation of the basic method, the 'extended sib-pair' method (ESPA), uses information on marker allele frequencies in the population to complete missing data and allows the method to be used in extended pedigrees. In general, the power to detect linkage is much lower than in the LOD score method, although this may not be the case in schizophrenia, where there are many unknown parameters.

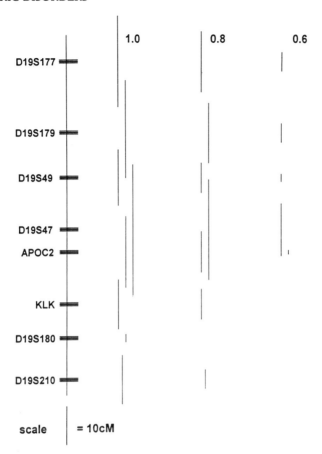

Fig. 9.4. Exclusion data for schizophrenia on chromosome 19. Regions of exclusion are defined using the conventional criteria of a LOD score of −2. This defines regions either side of each marker locus, illustrated by vertical lines. Here we demonstrate the effect of three levels of heterogeneity on our confidence to exclude these regions. Note how difficult it is to exclude the possibility of a gene of major effect in schizophrenia once heterogeneity is taken into account

$\alpha + 1.0$: assumes all pedigrees share a single disease locus on chromosome 19
$\alpha + 0.8$: assumes 80% of pedigrees share a single disease locus
$\alpha + 0.6$: assumes 60% of pedigrees share a single locus

ASSOCIATION STUDIES

While linkage analysis enables the detection of genes of major effect, it will not detect genes of small effect which may be neither necessary nor sufficient to cause the disorder. In this situation a single gene may not be linked to the disease phenotype, although it may show allelic association with that disease.

The ability of association studies to detect genes of small effect has made these studies popular in psychiatry. There are, however, several drawbacks including the problems of diagnosis and the question of comparability of patient populations from different centres. Applying strict operational definitions of illness can ensure that a patient sample is diagnostically homogeneous but does not avoid the possibility that there is aetiological heterogeneity.

The selection of controls provides a further pitfall, since there may be a section of the population in which a particular marker and a certain disorder are common without there being any casual relationship. For example, HLA BW16 is more common in Ashkenazi Jews, so that an excess of Jewish patients in an affected sample could lead to the false conclusion that an association exists between the disorder and that antigen. This effect is termed stratification. It is, however, possible to use parental genotypes, by comparing the frequencies of the parental alleles not inherited by an affected individual with the alleles which are inherited, thus providing a perfectly matched internal control. For this method it is necessary to genotype both parents of each affected subject.

Another problem is the statistical handling of results. Since the prior probability of obtaining a true association is extremely remote, the conventional level of statistical significance ($p < 0.05$) cannot be accepted. In addition, account must be made for the use of multiple markers in these studies. A conservative correction is to multiply the obtained p value by the number of markers tested.

The task of carrying out a systematic search for association throughout the entire genome would involve the use of approximately 3000 markers, each showing linkage disequilibrium with its neighbour. While this is at present an impossibly large amount of work, it is likely to become increasingly feasible as automated technology is developed. For the moment, the best strategy is probably to focus upon markers that are close to or within candidate genes.

An alternative approach is to look for variations which affect protein structure and expression (VAPSEs), rather than relying upon functionally silent polymorphisms (Sobell et al., 1992). The problem here is that we understand little about the pathophysiology of schizophrenia and plausible candidate genes are few and far between. In conclusion, it is probably fair to say that linkage and association studies have complementary roles to play in molecular genetic research of schizophrenia, at least until definitive findings point the direction for future studies (Owen and McGuffin, 1993).

LINKAGE STUDIES IN SCHIZOPHRENIA

Early linkage studies used classical markers such as HLA antigens, blood group types, serum protein polymorphisms and red cell enzymes. Studies with HLA showed promise when a maximum LOD score of 2.57 was obtained at a recombination fraction of 0.15 between a broadly defined phenotype 'schizotaxia' (similar to the concept of schizophrenia spectrum disease) and HLA. The analysis assumed an autosomal dominant mode of transmission. These findings were not, however, replicated by a further four linkage studies and a 'model-free' sibling-pair analysis which showed substantial evidence against linkage. Studies with other classical markers have not provided evidence of linkage. Taken together the exclusion of linkage to these markers has ruled out about 6% of the genome.

The advent of DNA markers available throughout the genome has now opened the way for more extensive linkage studies. Several strategies are possible for this task. First, the identification of candidate genes such as neural cell adhesion molecules, neuroreceptor proteins and enzymes involved in the metabolism of neurotransmitters allows direct tests for linkage using genetic markers within the candidate genes themselves or adjacent polymorphisms. Secondly, the segregation of a chromosomal abnormality such as a translocation or deletion with the disorder may provide clues to the localization of disease genes. Thirdly, the sytematic typing of highly polymorphic markers such as microsatellite repeats at regular intervals will allow a complete genome search.

Great excitement resulted from high LOD scores obtained between markers in 5q11–13 and schizophrenia. This region had been identified by the report of a Canadian family of oriental origin in which a young schizophrenic and his schizophrenic uncle both had a partial 5q trisomy due to an unbalanced translocation. The positive findings were based upon two British and five Icelandic pedigrees. LOD scores of 2.45 were obtained when cases of schizophrenia alone were defined as affected, 4.33 when the phenotype was broadened to include mild schizophrenic-like syndromes and 6.49 when other psychiatric diagnosis (major and minor affective disorders, phobic disorder and alcoholism) were included. The initial optimism that these results produced has been severely tempered by the inability of several subsequent studies to confirm them. It is not clear why these discrepancies exist but there are several possibilities. The initial response was that the finding of positive linkage in the first study was correct and that failure of other studies to demonstrate linkage reflected non-allelic heterogeneity. However, this now seems unlikely since no evidence for linkage emerged when pedigrees studied by other groups were subjected to heterogeneity analysis (McGuffin et al., 1990). Follow-up of 'linked' pedigrees and further typing with more informative markers has confirmed that the original results were falsely positive (Mankoo et al., 1991).

The long arm of chromosome 11 has been another region of interest, with a number of separate findings indicating the possible involvement of a gene or genes. Three families were independently ascertained in which psychiatric illness appeared to be segregating with a balanced translocation of 11q. A family was described in which schizophreniform psychosis is segregating with tyrosinase-negative oculocutaneous albinism—a disease whose gene maps to 11q14–21. The dopamine D2 receptor has been implicated in the pathogenesis of schizophrenia, since dopamine receptors have been implicated as the main site of action of antipsychotic drugs. The human dopamine D2 receptor gene is located on 11q21–23. In addition, the gene for porphobilinogen deaminase (PBG-D) is located on 11q23–qter. This is of interest since individuals with acute intermittent porphyria resulting from mutations in this gene may present with psychiatric symptoms, including psychosis. Further interest in this gene has arisen from a report of allelic association between schizophrenia and one of the alleles of an MspI polymorphism at this locus, although this finding was not replicated in a subsequent study carried out in Wales.

Unfortunately, the obvious interest in this chromosome has not been followed by any reports of positive linkage findings. Study of a large Swedish pedigree using markers close to the dopamine D2 receptor effectively excluded the region. The entire long arm of chromosome 11 has been screened with a set of highly polymorphic markers at approximately 10 cM intervals in a set of 24 pedigrees, with no evidence of positive linkage and evidence of exclusion under a wide range of genetic parameters. Since then two further studies have come to similar conclusions.

On the basis of an excess of sex chromosome aneuploidies among patients with schizophrenia and reports that schizophrenic siblings are more often of the same sex than of opposite sex, a genetic locus for schizophrenia within the pseudoautosomal region of the X chromosome has been proposed (Crow, 1988). This hypothesis was tested using a sibling-pair strategy, with a highly polymorphic marker from the telomeric end of this region. This study found that affected siblings shared alleles more often than expected under random segregation ($p < 0.05$). Following this, the finding was replicated in a sample of French families ($p < 0.05$), but not in a sample of American families nor in a sample of Welsh pedigrees using both sibling-pair and LOD score methods.

Chromosome 21 has also been examined using markers spanning the entire long arm, with no evidence of linkage and exclusion under a wide range of genetic and diagnostic models. In addition, this study used single-stranded conformational polymorphism analysis (SSCP) to look for mutations within part of the coding sequence of the APP gene, since it had been reported that a mutation within this gene had occurred in an individual with schizophrenia. No sequence variation was detected in a sample of 68 unrelated schizophrenics, suggesting that the reported association was almost certainly a chance finding. This approach of looking directly for gene mutations within the coding and controlling regions of candidate genes is an important strategy and will be discussed further in the section on association studies.

ASSOCIATION STUDIES IN SCHIZOPHRENIA

Early association studies in schizophrenia used classical markers such as the ABO, other blood groups and the HLA system. The results of these studies are inconsistent and when considered overall no clear evidence for association is apparent. The most likely explanations for these inconsistencies—multiple testing and stratification—have already been discussed.

The most interesting finding was of possible association between a subtype of schizophrenia, paranoid schizophrenia, and HLA-A9 in seven out of nine studies. Combining the data and applying a correction for multiple testing gave a p value of 0.0003. However, several conflicting findings have been found. Two groups found A9 to be decreased in paranoid schizophrenia compared with controls, while others using samples from the same countries found A9 to be increased. A9 consists of two subspecificities: AW23 and AW24. Two studies suggested an association with AW23, while others found a stronger relationship with AW24. A more recent study found no association between paranoid schizophrenia and either AW23 or AW24. Another recent study of 33 pedigrees collected in France failed to find evidence of linkage between HLA and the schizophrenic phenotype. In addition, this study reported association findings using pooled data from six independent studies and were unable to show a significant excess of HLA-A9 in the affected group, but they did not examine the paranoid subtype on its own.

The candidate gene approach in schizophrenia is perhaps best exemplified by studies of dopamine receptor genes. Until recently it was thought that there were only two dopamine receptor subtypes: D1 and D2. These differ in ligand binding and function since D1 suppresses adenylate cyclase, whereas D2 stimulates it. However, homology cloning has resulted in a further three dopamine receptor genes being isolated, namely the dopamine D3, D4 and D5 receptor genes. D3 and D4 have a similar structure to D2 and bind D2 selective ligands, while D5 is similar in structure and ligand binding to D1.

Interest was focused on D2-like receptors, since the therapeutic effects of antipsychotics are thought to be mediated by these—a view which has now been modified to include the D3 and D4 receptor genes. D3 is of particular interest since its expression is restricted to brain limbic regions implicated in schizophrenia. Furthermore, unlike D1 and D2, the expression of the D3 receptor gene appears to be increased by the use of both older 'typical' antipsychotics and more recently developed drugs which have more specific antipsychotic activity.

Recently, two groups from Wales and France independently carried out association studies of a BalI polymorphism within the dopamine D3 receptor gene, which gives rise to a glycine to leucine substitution (Crocq et al., 1992). In both studies, more patients than controls were homozygotes of either type ($p = 0.005$, $p = 0.008$). Pooling of the data gave a highly significant result ($p = 0.0001$) with a relative risk of schizophrenia in homozygotes of 2.61 (95%

confidence intervals 1.60–4.26). The Welsh study has been extended and results suggest that this effect may be stronger in patients with a family history of mental illness, in males and in those who respond better to medication.

A difference approach is to look for variations within candidate genes that may be associated with drug responsiveness. Of recent interest has been the drug clozapine, which is widely reported to reduce symptoms of schizophrenia which are unresponsive to other antipsychotic medication. However, some patients remain unresponsive to clozapine and it has been suggested that response to this drug may be related to variation of the D4 receptor gene (DRD4). This view arises from the observations that clozapine has a particularly high affinity for D4 receptors and that the D4 receptor shows an unusually high degree of variation. The hypothesis is currently being tested by studying the relationship between particular allelic variants of the gene in patients who are good clozapine responders and those who are clozapine-unresponsive. A recent study of DRD4 variation in a sample unselected for drug responsiveness shows no evidence of association.

MANIC DEPRESSION

CLINICAL SYNDROME

Manic-depression or bipolar affective disorder has a lifetime prevalance of around 1%, with an incidence of between 9 and 15 per 100 000 in men and 7 and 30 per 100 000 in women. It is an illness which often affects young people, with an onset in the second decade. The term 'manic-depression' comes from the two opposing syndromes of depression and mania which occur in these patients. Depressive episodes occurring alone are relatively common so it is the presence of at least one manic episode which distinguishes the disorder from other affective disorders. In mania the central feature is elevation of mood which is normally accompanied by increased levels of activity and subjective feelings of well-being. The patient may be cheerful, bubbling with enthusiasm and optimistic but the mood is often highly labile, and rapid changes to irritability, anger and subjective discomfort are common. Overactivity can be marked, with the patient unable to keep still and unable to sleep. Social disruption often occurs as a result of reckless or disinhibited behaviour for which the patient can show no insight at the time. Overspending, unrealistic business schemes and excessive sexual activity are not uncommon and are clearly of great concern. Speech is often rapid or pressured and may be accompanied by a disorder of thought called flight of ideas, in which thoughts rapidly move between different topics only loosely related or related by rhyme, puns and jargon phrases. Delusions and auditory hallucination may occur which are usually congruent with the mood. The presence of thought disorder, delusions and hallucinations can make the differential diagnosis from other psychoses, in particular schizophrenia, very difficult and a matter of fine judgement. The course of the illness tends to be different from that of schizophrenia, being episodic with good recovery between episodes, whereas schizophrenia tends to follow a more chronic course.

FAMILY, TWIN AND ADOPTION STUDIES

As in schizophrenia, family, twin and adoption studies point to a substantial genetic influence on susceptibility (see Craddock and McGuffin, 1993, for review). Studies of affected individuals and their first-degree relatives show that manic-depression and schizophrenia in general breed true, and only rarely occur together in the same family. Suggestions that schizoaffective disorders represent an intermediate form along a continuum of psychosis with manic-depression and schizophrenia at the extreme ends are contradicted by the relative rarity of twin pairs containing both these diagnoses. Kraepelin in 1922 was among the first to note that heritable factors were apparent in 80% of his patients affected by manic-depressive insanity. In 1959 Leonhard made the important distinction between bipolar disorder (BP) with episodes of both mania and depression, and unipolar disorder (UP) with episodes of depression alone. This distinction was first used in family studies in 1966 by two authors: Perris and Angst. Perris found that there was striking evidence of homotypia (breeding true), but the findings of Angst did not confirm this. Angst found that among the relatives of UP probands there was an increase in the rates of UP disorder only, whereas among the relatives of BP probands both UP and BP disorders were raised. Further studies have tended to confirm the findings of Angst. A review of 12 studies estimated that on average 7.8% of first-degree relatives of BP patients had bipolar disorder and 11.4% had unipolar disorder. In comparison 9.1% of first-degree relatives of UP patients had UP disorder but only 0.6% had BP disorder—a figure comparable with that found in the general population.

Twin studies have consistently shown a higher concordance rate between monozygotic (MZ) than dizygotic (DZ) twins. In accordance with family data the genetic influence appears to be greater for BP than in UP disorder. In a Danish twin study, the MZ : DZ ratio in BP disorder was 3.5 : 1 compared with 2 : 1 in UP disorder. Using a broad definition of BP disorder, concordance rates among MZ twins reach almost 100% in some series.

There have been fewer adoption studies in affective disorder than in schizophrenia. The study of Mendlewicz and Rainer in 1977 showed that 28% of the biological parents of BP adoptees had affective illness, compared with 12% in the adopting parents. The authors also noted that 26% of parents of non-adopted BP probands were affected, which suggests that their findings cannot be due to increased rates of adoption amongst the offspring of bipolar parents.

THE MODE OF TRANSMISSION

While the importance of genetic factors is beyond doubt, the precise mode of transmission is unknown. As described for schizophrenia, it has not been possible to distinguish between single-gene, multifactorial and mixed models, and it is perhaps likely that there is genetic and aetiological heterogeneity. Likewise, there are families heavily loaded with affected individuals, in which inheritance appears to be dominant, and these are the most likely to be segregating major genes. However, twin and family data suggest that BP disorder may have a particularly high heritability and is perhaps more likely to involve a single major gene than is schizophrenia.

GENETIC MARKER STUDIES

Early association studies were carried out with classical markers. Significant associations with the ABO blood group and the HLA system were reported. However, the results of these studies are conflicting and no consistent picture emerged, so it is unlikely that any true association was found. The particular difficulties in association studies of diagnostic uncertainty, the selection of controls and multiple testing have been discussed already and have doubtless played a part here.

The first linkage studies also used classical markers. A great deal of excitement arose following a report of linkage to HLA and claims that a major gene locus had been found on chromosome 6. This finding was, however, followed by several failures to replicate it. A recent reanalysis of several studies with HLA markers suggests that linkage with affective disorder can be excluded to a distance of 20–25 cM either side of this region.

There has been great interest in the short arm of chromosome 11 since a group studying the Old Order Amish in Pennsylvania reported linkage between bipolar affective disorder and the Harvey-ras-1 (hRas) and insulin (Ins) loci which both map to 11p15. They obtained a maximum multipoint LOD score of 4.9 under a dominant model for the mode of transmission. However, this finding was not replicated in other families by several other groups. As in schizophrenia these conflicting findings were initially thought to provide evidence of genetic heterogeneity. This seemed especially plausible since the Old Order Amish are a large inbreeding population and are genetically isolated from the pedigrees used by other groups. However, the plausibility of this explanation was undermined when follow-up of the Amish families failed to confirm the initial linkage findings. The maximum LOD score for hRas dropped to 1.75 after two previously unaffected individuals became unwell and 12 additional family members were typed. The original core pedigree was also extended with the addition of new family branches, and this resulted in exclusion of a region 15 cM either side of the original markers. This latter finding has led some to propose that there is more than one major gene, each segregating in different family branches of the Amish.

Interest in this chromosomal region has been maintained by a reported association between bipolar disorder and a polymorphism at the tyrosine hydroxylase locus, which also maps to 11p, close to Ins and hRas. Tyrosine hydroxylase is of particular interest because it is the rate-limiting step in catecholamine synthesis and is therefore a strong candidate gene in manic-depressive illness (Fig. 9.5). The group reporting this association were unable to find linkage in a sample of extended families, suggesting that a gene at this site may only contribute a small effect to the liability to develop bipolar affective disorder. If this is so, a mutation of tyrosinase hydroxylase could

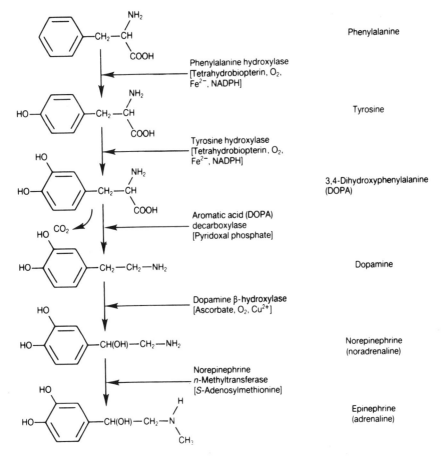

Fig. 9.5. Synthesis of catecholamines from phenylalanine. Tyrosine hydroxylase is the rate-limiting step. From Smith CUM, *Elements of Molecular Neurobiology*. Chichester: Wiley

contribute to the risk of developing bipolar disorder in the Amish. The positive linkage score could arise if another gene or genes having an additive of interactive effect were 'fixed' in part of the pedigree, while other branches show variation at these loci.

THE DEMENTIAS

Dementia is a common clinical syndrome characterized by global impairments to cognitive function. The prevalence of dementia is extremely high in old age, being around 5% in those over the age of 65 and rising to 20% for those over 90. Approximately 50% of all cases are due to Alzheimer's disease (AD), 20% to multi-infarct dementia (MID), 20% to mixed AD/MID and the remainder to various other diseases. In recent years there have been major advances in our understanding of the genetics of Alzheimer's disease and the much rarer spongiform encephalopathies. On the other hand, MID remains poorly understood. The difficulties in defining MID and separating it clinically from AD and other dementias have meant that few family studies have been conducted, but these have consistently shown an increased risk to first-degree relatives. While certain predisposing factors such as blood pressure and arterial disease are under some degree of genetic control, there is little direct evidence to help define the degree of genetic influence.

ALZHEIMER'S DISEASE

AD is a common disease affecting approximately 1% of the population at age 65 and 10–20% in the ninth decade. It is a progressive neurodegenerative disorder which culminates in a marked dementia. There is a characteristic neuropathology, with the appearance of large numbers of senile plaques and neurofibrillary tangles (Fig. 9.6). Senile plaques consist of an extracellular protein fragment called β-amyloid protein. This arises from a much larger protein, the β-amyloid precursor protein (APP). In 'mature' plaques the central core of β-amyloid is surrounded by abnormal neurites and glial cells. There are also abundant, amorphous, non-filamentous deposits of β-amyloid. These 'diffuse' plaques contain little or no degenerating neurites or reactive glial cells and are likely to be precursors of the mature plaques.

Neurofibrillary tangles are dense bundles of abnormal fibres or paired helical filaments (PHFs) found in the cytoplasm of certain neurones. They are composed, at least in part, of an altered form of the microtubule-associated protein, tau. The remaining constituents of PHFs are unknown. In particular, the question of whether β-amyloid is an intrinsic component of PHFs is as yet unanswered. Both senile plaques and neurofibrillary tangles are found in a number of other chronic cerebral disorders and in increasing amounts in normal undemented people as age increases. In AD they are more numerous

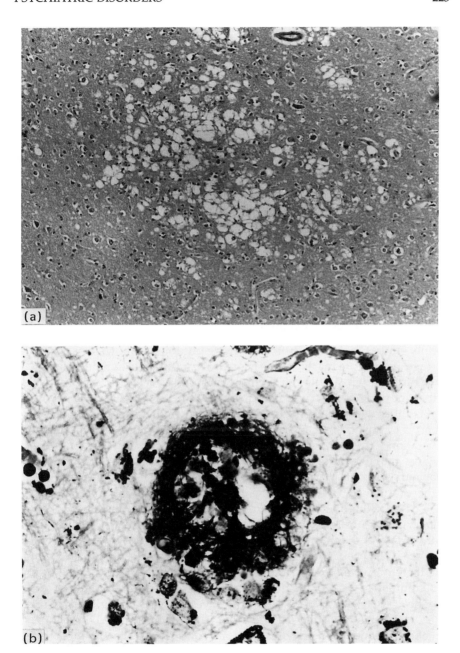

Fig. 9.6. (a) Silver-stained preparation of cortical brain tissue showing gross Alzheimer's disease pathology. The amorphous darkly stained deposits are amyloid plaques in different stages of progression and the darkly stained rods are neurofibrillary tangles. (b) Preparation of cortical brain tissue showing the characteristic vacuolar degeneration found in prion diseases

and widespread than in the undemented brain. AD can be viewed, neuro-pathologically at least, as accentuated and accelerated normal ageing.

Several important epidemiological observations have been made which help our understanding of genetic factors in AD (see Owen, 1994, for review). The main risk factors identified are increasing age and the presence of a family history. Although most cases of AD are sporadic, multiplex families do occur and over 100 families have been reported in which AD appears to segregate as an autosomal dominant disorder of early onset. The term 'familial Alzheimer's disease' (FAD) has been used to describe cases from these pedigrees, although there are no consistent clinical or neuropathologi-cal features which distinguish them from 'sporadic' cases. Clear evidence of autosomal dominant transmission is present in only between 0.1% and 1% of all cases. The other well-established fact is that the majority of individuals with Down's syndrome (trisomy 21) who reach the fourth decade of life develop the characteristic neuropathology of AD. This relationship between AD and Down's syndrome suggests that a locus or loci on chromosome 21 may be important in the aetiology of AD.

Family, twin and adoption studies

Family studies in AD are difficult to carry out for several reasons. First, the late age of onset means that some genetically susceptible individuals will die prior to developing the illness. Second, a proportion of individuals unaffected at the time of assessment will go on to become affected. Finally, it is unusual to find families with living affected individuals in more than one generation, so that it is rarely possible to obtain DNA from the members of more than one generation.

Early family studies found increased rates of illness in relatives of probands with AD. However, for early-onset disease, the rates were much higher, so that the cumulative incidence of AD in siblings and children of such cases with an affected parent approached 50% by age 85–90.

Later studies differed from those of early investigators in the use of far more stringent diagnostic criteria, although autopsy evidence was seldom available. The probands in all these studies were unselected for age of onset and therefore consisted predominantly of senile cases. In all of these studies the cumulative risk of dementia in first-degree relatives did not differ significantly from 50% by age 85–90 years. These results have been interpreted as supporting the view that the aetiology of AD has a major genetic component which shows autosomal dominant transmission with age-dependent penetrance.

Interpretation of these results should be made with some caution, as certain causes of bias may exist. The family history method used in these studies may well have included cases with other dementias, such as multi-infarct dementia, thus overestimating the number of cases with true AD. In addition, it is possible that ascertainment bias operated to increase the number of cases with high familial loading. To overcome these possible sources of error, Farrer and colleagues (1989) identified all those cases in their series who were likely to have been referred because their illness appeared familial. When these cases were excluded from the analysis, the cumulative incidence of AD in first-degree relatives was between 24% and 39%, compared with 36–52% when they were included. They also showed that the cumulative risk of disease in the first-degree relatives of early-onset cases was not higher than in the first-degree relatives of late-onset cases. However, the relatives of early-onset cases had significantly early ages of onset themselves. It is therefore possible that it is the age of onset in relatives that is under genetic control. This is further suggested by the observation that intrafamilial variation in age of onset in FAD is low.

To date there is no convincing evidence from twin and adoption studies implicating genetic factors in the familial aggregation of AD. No adoption studies have been carried out. An early twin study showed an MZ:DZ concordance ratio of 43:8. However, the cases included suffered from 'senile psychosis' and it is unlikely they all suffered from AD. A more recent study with modern diagnostic criteria failed to find a significant difference in the MZ and DZ concordance rates, being 40% and 41% respectively, but cases from this study were not systematically ascertained or followed up.

Mode of transmission

As discussed above, there are several lines of evidence suggesting autosomal dominant inheritance, in at least a proportion of cases of AD. Families with FAD appear to show autosomal dominant transmission, and as we shall see later the existence of such families is now proven. In addition, the evidence from recent family studies suggests that AD occurring in individuals unselected for age or family history may also follow autosomal inheritance with age-dependent penetrance. However, most cases of AD do not show striking patterns of familial inheritance, there are many reports of discordant MZ twins, and the family study of Farrer and colleagues suggests that the cumulative risk of 50% described in other studies may be an overestimate. These findings are more in accord with a complex pattern of inheritance in which a major gene, or genes, operate against a polygenic or multifactorial background. A recent complex segregation analysis is in accord with this hypothesis, suggesting that while a major autosomal dominant gene is likely to exist, it is unlikely to be acting alone, but rather it is acting against a polygenic background. On the other hand, the possibilities of multifactorial inheritance only, a recessive inheritance and no genetic susceptibility were all strongly rejected. Farrer and his colleagues point out that their best-fit model supports the existence of genetic heterogeneity. According to this view, while in some families the disease is primarily caused by a single gene transmitted in an autosomal dominant manner, other cases of AD, perhaps the majority, result from polygenic multifactorial inheritance.

Molecular genetics of Alzheimer's disease

The first report of positive linkage was from St George-Hyslop and colleagues (1987), who found linkage between AD and several DNA markers that mapped to the long arm of chromosome 21 (loci D21S1/S11 and D21S16) in four pedigrees multiply affected by early-onset disease. At about the same time as this initial report the gene for APP was cloned and localized to the same region of chromosome 21. It was further reported that several sporadic cases had duplication of the APP gene. It therefore seemed likely that a mutation in the APP gene, leading to overexpression, might be the primary genetic defect in FAD. Furthermore, it was suggested that extra copies of the APP gene led to overexpression in sporadic AD and the changes in Down's syndrome. However, this simple account of events was confounded by two reports of recombination between the APP gene and AD in several multiply affected families. In addition a number of studies were unable to find a duplication of the APP gene in sporadic cases.

Two groups did go on to replicate the finding of chromosome 21 linkage in early-onset FAD. These studies placed the locus close to D21S16, which is a considerable genetic distance from APP. Other studies have not found linkage to chromosome 21 markers. In a study of seven families of Volga German decent, significant evidence against linkage to D21S1/S11 was found. These North American pedigrees consist of early-onset cases where ancestors can be traced back to a small ethnic population on the banks of the Volga River. In addition, a study of small families with predominantly late onset was unable to detect linkage to chromosome 21 markers.

St George-Hyslop and colleagues extended their study to include 48 families of both early and late onset. Using a number of markers from chromosome 21 their analyses supported the earlier finding of linkage. However, they obtained evidence of heterogeneity by dividing the sample into early and late-onset families, demonstrating linkage in the early-onset families but not the late-onset families. However, weak evidence of chromosome 21 linkage in late-onset families has been provided, in a study using a model-free method of analysis, as well as rather more convincing evidence of chromosome 19 linkage using both model-free and LOD score methods.

The results of the linkage studies described above all point towards aetiological and genetic heterogeneity in AD. There is clear evidence of linkage to markers on chromosome 21 in some, but not all, early-onset cases of FAD. The results from AD of late onset are more difficult to interpret because the mode of transmission is not well understood. Unlike the early-onset FAD cases it is not certain that clustering of late-onset AD follows Mendelian segregation patterns. While this might be so in a proportion of late-onset families it is likely that some result from more complex modes of inheritance. Despite this, there is currently suggestive evidence of loci involved in late-onset AD on chromosome 21 and chromosome 19. It seems reasonable to assume that the former is APP though the precise molecular pathology will probably be different.

The demonstration of genetic heterogeneity among early-onset FAD has implications for the interpretation of linkage results. If a sample of families used in a linkage analysis contains two types—one linked to the marker being studied and the other unlinked—then the estimate of θ (recombination fraction) is likely to be increased. In other words, the genetic distance between the linked disease locus and the marker will be overestimated. This means that linkage to the APP gene could still exist in chromosome 21 linked families despite results which pointed to a locus some distance from that gene.

Further interest in this possibility was fuelled by the finding that hereditary cerebral haemorrhage with amyloidosis of the Dutch type (HCHWA-Dutch) is caused by a mutation in exon 17 of the APP gene. This is a rare disease found in two Dutch families in which cerebral haemorrhage occurs due to extensive deposition of β-amyloid in cerebral blood vessels. The fact that β-amyloid angiopathy occurs in both HCHWA-Dutch and AD led Goate and colleagues (1991) to sequence exon 17 of the APP gene in a British family with FAD which was large enough to show significant evidence for linkage to chromosome 21 markers. Sequencing revealed a C to T mutation which led to a valine to isoleucine substitution at amino acid 717. The same mutation was subsequently found in an American family. Following this the mutation was also found by other workers in two Japanese and one British family. Other mutations in codon 717 have also been found, resulting in a valine to glycine substitution and a valine to phenylalanine substitution. To date these mutations have not been found in several hundred unaffected individuals, supporting the conclusion that they are pathogenic. These mutations are, however, rare, occurring in only some cases of FAD and have not been found in 'sporadic' or late-onset cases. Fig. 9.7 illustrates the position of this and other mutations found in APP.

In recent years there have been two additional major advances, confirming the complexity of genetic inheritance of this disorder. First, a second locus on chromosome 14 has been detected which like APP is linked to early-onset FAD and segregates in an autosomal dominant fashion. This finding has been replicated by more than four groups and work is now being undertaken to detect the defective gene.

More surprising is the very recent demonstration that the E4 allele of apolipoprotein E (apoE) is in association with both familial and sporadic late-onset AD (Saunders *et al.*, 1993). The frequency of this allele was found to be 52% in affected members of late-onset multiply affected families and 40% in a series of post-mortem confirmed cases of sporadic late-onset cases. This compares with a rate of only 15% in control samples. The initial findings have

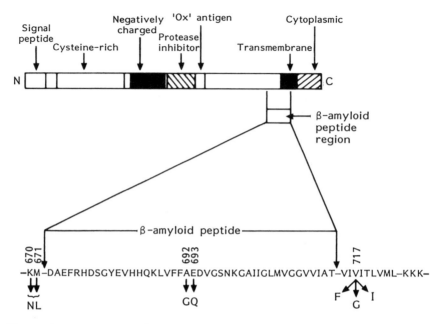

Fig. 9.7. Mutations in amyloid precursor protein (APP)

been replicated by several groups. It is important to realize that approximately 40–50% of those with AD do not have the E4 allele and many individuals with E4 will not develop the disease. However, presence of this allele confers an increased risk for the disorder. The degree of risk and severity of the disorder appear to be related to the number of E4 alleles since individuals homozygous for E4 are at higher risk and tend to have an earlier age of onset than heterozygotes. It is notable that within multiply affected families over 90% of E4 homozygotes develop AD before the age of 80 and this may also be true in those unselected for family history.

ApoE is produced in many tissues, including liver and nerve, and is known to function in the transport and distribution of lipids by interaction with lipoprotein receptors. While it is possible that apoE is merely acting as a marker for some other genetic mutation with which it is in linkage disequilibrium, there are a number of findings which point to its direct involvement in the pathogenesis of AD. It is known to be a neuronal stress protein since its synthesis is greatly increased following injury. ApoE is a constituent of amyloid plaques and has also been found in neurofibrillary tangles. In a recent study it was demonstrated that apoE binds avidly to synthetic β-amyloid and this binding is greater for the E4 form than the E3 form. Another study has shown that AD brains from individuals with either one or two E4 alleles have increased amyloid deposition compared with AD brains from individuals that do not have this allele. Together these findings suggest that the E4 allele may act by facilitating the deposition of amyloid deposits.

In conclusion, it is clear that major inroads are being made into our understanding of AD. The identification of mutations within the APP gene, while rare, points to the central role of β-amyloid in the pathogenesis of this disease. Further experiments with transgenic mice are likely to reveal some of the detailed mechanisms involved. Recent work has located another major locus on chromosome 14 and it is likely that other major loci exist in a few affected families. The chromosome 14 gene has not been isolated yet and its function is therefore unknown, but it is perhaps likely that this and other genes will be involved in the metabolism and degradation of β-amyloid. The situation with late-onset AD has always been thought to be more complex and the finding of an association with the E4 allele of the apoE gene confirms this. Further work is needed to uncover other factors involved in these common, often sporadic cases. It is likely that whatever these determinants of AD turn out to be they are likely to act, at least in part, through the abnormal processing of β-amyloid and its abnormal deposition.

PRION DISEASES

The spongiform encephalopathies or transmissible dementias are a group of neurodegenerative disorders which in humans consist of Creutzfeldt–Jakob disease (CJD), Gerstmann–Straussler syndrome (GSS) and kuru. They are highly unusual in that they are transmissible and also in some cases inherited. Variants of these disorders occur within other species and some can apparently be transmitted by ingesting infected material. A so-called 'prion' protein has been identified which is central to the pathogenesis of these diseases and seems likely to be responsible for the infective process without the involvement of nucleic acids. Since the discovery of prions it has been possible to examine a wide range of dementia syndromes for their presence and it is clear that 'prion disease' has a wider range of expression than previously thought.

In 1950 the similarities between CJD and sheep scrapie were first noted. Since then identical diseases have been found in many other species, the most notorious being bovine spongiform encephalopathy (BSE or 'mad cow disease'). Although the human spongiform encephalopathies are rare, great interest in them was stimulated when it was shown that they could be transmissible by inoculation to other species and from person to person following surgical procedures. Early research in this field was aimed at identifying the infectious agent, which was originally presumed to be a slow virus.

Typically, CJD is a pre-senile dementia without a family history, occurring in the sixth decade of life and progressing rapidly to death. There is an atypical form of CJD which is familial, with an earlier age of onset. The familial forms are less transmissible to primates (64% of cases compared with 90% of sporadics) and account for 6–15% of all CJD cases. The neuro-pathological appearance is of prominent spongiform changes in the cerebral cortex, due to cystic dilation of neurones, focal necrosis and massive neuronal loss (Fig. 9.6). In 15% of cases the brains contain amyloid plaques, which although morphologically similar to senile plaques in AD are formed from prion protein, rather than β-amyloid. Clinical diagnosis of the typical case is relatively straightforward, but the occurrence of clinical variants means that definite diagnosis requires the demonstration of the typical neuropathology or better still its transmissibility to primates. In addition, it is now possible to use genetic diagnosis for known mutations. The differential diagnosis includes all forms of ataxia and dementia.

GSS differs from CJD in several respects. It occurs less commonly than CJD and is transmitted as an autosomal dominant disorder. The age of onset is earlier, with a mean age of death about 50. The pathological findings are similar to those found in CJD except that amyloid plaques are invariably found in the cerebellum. It usually causes mild dementia and cerebellar signs are usually prominent. Kuru is a similar disease to GSS but is restricted to the Fore tribe of New Guinea, where it was transmitted by the practice of endocannibalism. It has now virtually disappeared.

Evidence that the infective agent in spongiform encephalopathies is unusual came from the observation that in kuru and scrapie the infectivity of brain tissue was not removed by procedures which destroy nucleic acids, such as treatment with traditional disinfectants and preservatives. The infective agent also passes through fine-pored filters which do not allow the passage of larger agents such as viruses. On the other hand, procedures which destroy proteins do remove the infectivity of these tissues. In cellular fractionation experiments infectivity was related to a particular protein fraction of protease-resistant prion protein (PrP). No other proteins or nucleic acids have been shown to be associated with infectivity, thus establishing the central role of prion protein. It was further shown that antibodies raised to PrP in scrapie-affected mice show immunological reactivity to similar proteins found in the affected brains from other species, including the human brain.

In the mid-1980s, the human PrP gene was identified and localized to chromosome 20. Gene expression occurs in neurones and glia throughout life and its function appears to be that of a 'housekeeping' gene necessary for the normal functioning of these cells. Studies of familial CJD and GSS have revealed that mutations of this gene segregate with the disease in a number of families. These findings strongly suggest, but do not prove, that these mutations are pathogenic.

The first mutation of the prion protein gene to be characterized was reported in 1989. This consisted of an insertion of an extra 144 bp at codon 52 of the gene in affected members of a CJD family. This mutation has since been found in a family clinically diagnosed as having early-onset AD, and five out of 101 atypical dementias. A single base pair mutation which resulted in a proline to leucine change (codon 102) was reported in a GSS pedigree and subsequently this mutation has been found in other families with GSS and also CJD. Other reported mutations include single base pair changes at codons 189 and 200, both resulting in amino acid changes. These mutations have not been found in all cases of transmissible dementia and it is assumed that there are other mutations yet to be identified or that post-translational changes to the protein have occurred.

Further evidence of the importance of the prion protein came from studies on incubation times for scrapie in infected mice. It had been known that incubation time was affected both by the host strain and the source of the infectious agent. Linkage analysis was carried out to determine the genetic nature of this 'incubation time gene' and it was found to be linked to the PrP gene. Subsequently specific mutations in the gene were found to relate to different incubation periods for the disease. Furthermore, prion genes were introduced into transgenic mice and showed that different mutations of the PrP gene resulted in different incubation times, different pathological features and altered susceptibility to the infection.

Despite identification of both the disease gene and its product, it is not yet clear how the abnormal PrP causes the pathological process. While its heritability is easily understood by the segregation of mutant forms through successive generations, it is not yet clear whether infectivity is due to abnormal PrP itself. Several recent findings have, however, led to new hypotheses of disease pathogenesis. Animal experiments show that cross-species infection is less efficient, both in terms of the number of animals infected and the duration of incubation time, than infection within species, leading to the proposal of a 'species barrier'. Recent transgenic experiments by Prusiner and colleagues (See Prusiner, 1994, for review) demonstrate that it is likely to be the PrP gene product which is the pathogenic agent and that the host PrP is the major determinant of the species barrier. Mice were transfected with a PrP gene containing a mutation homologous to that found in two families with GSS. These animals developed spongiform encephalopathy and isolates derived from them were used successfully to infect other mice. Thus an infectious agent had apparently been produced by the simple substitution of a single amino acid in PrP. In subsequent experiments, mice that were transfected with a hamster PrP gene became more susceptible to infection from hamster PrP than mouse PrP and produced large amounts of hamster PrP. On the other hand, when these hamsters were infected with mouse PrP they produced large amounts of mouse PrP.

The importance of the interaction between infecting PrP and host encoded PrP is further demonstrated in a series of studies of human spongiform encephalopathies. In the general population there is a non-pathogenic allelic variant at codon 129; 70% have a methionine (M) in this position and 30%

have a valine (V). It has been shown that those with a VV genotype seemed to be particularly susceptible to iatrogenic infection. In one large British family with autosomal dominant transmitted spongiform encephalopathy, a pathogenic mutation has occurred on an M allele. Within this family those whose other allele was also M became ill under 40 years of age, whereas those whose other allele was the V variant had a later age of onset. This suggests that the interaction between the mutant PrP and the PrP produced by the normal allele is more pathogenic if an individual is homozygous at codon 129 rather than heterozygous. The greater sensitivity of PrP homozygotes to prion disease also holds for sporadic CJD, where the great majority of cases are homozygous for either allele at codon 129 compared with only about 50% of the general population.

Another important experiment was the construction of mice who had both copies of chromosomal PrP deleted. Unexpectedly, by the age of two years these mice had developed normally. In addition, they are resistant to prions and do not propagate scrapie infectivity. In contrast, mice who had a single chromosomal PrP deleted were not resistant to infection with mouse prions, but they showed prolonged incubation times following inoculation.

The central role for the interaction between infecting PrP and host-encoded PrP and possible mechanisms for this have been described (Fig. 9.8). One proposal supposes that there are two basic forms of PrP: a normal cellular isoform (PrP^c) and a pathogenic isoform (PrP^{Sc}). It is further proposed that

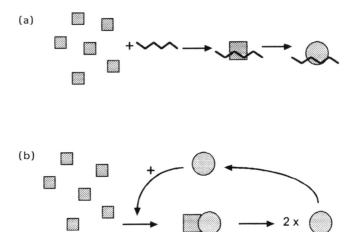

Fig. 9.8. Some possible mechanisms of prion replication. (a) Two-component prion model. Prions contain a putative, as yet unidentified, nucleic acid or another second component (solid, thick wavy line) that binds to PrP^c(squares) and stimulates conversion of PrP^{Sc} (circles). (b) One-component prion model—Prions devoid of nucleic acid. PrP^{Sc} binds PrP^c forming heterodimers that function as replication intermediates in the synthesis of PrP^{Sc}. Repeated cycles of this process result in an exponential increase in PrP^{Sc}. From Prusiner (1994), with permission

PrP exists both as a monomer and a dimer and that there is an equilibrium between these. If a dimer is formed from the pathogenic and the normal forms of PrP then the normal form will be converted to the pathogenic form. This would provide a mechanism for infectivity. In hereditary forms of the disease the PrP^{Sc} which is endogenously produced would spontaneously convert PrP^c to PrP^{Sc} and thus initiate the disease process and render PrP from that individual infectious. A further proposal is that PrP homodimers (identical molecules) are more stable that heterodimers. This would explain the species barrier, variability in age of onset in inherited spongiform encaphalopathies and the increased susceptibility in homozygotes to iatrogenically acquired disease.

While it is clear that abnormal PrP (PrP^{Sc}) is essential for the clinical appearance of the disease, there is still controversy over whether it is the infectious agent, or whether it is a by-product of viral infection. In a recent study scrapie-infected hamsters were treated with the antibiotic amphotericin B. This delays the development of clinical signs and accumulation in the brain of abnormal PrP^{Sc}. However, the infectivity of tissue from these amphotericin-treated hamsters did not differ from that of those not treated. This suggests that while PrP^{Sc} is necessary for development of scrapie, it may not be the infectious agent itself. The authors proposed that amphotericin B might modify a hypothetical receptor molecule for scrapie agent on target cells of the nervous system, or that it delays replication of the scrapie agent, or that it might interfere with the formation and accumulation of the pathological forms of PrP in the brain of infected animals. They further propose that these three possibilities may not be mutually exclusive if PrP is a scrapie agent receptor and PrP agent binding promotes the accumulation of PrP^{Sc} and of infectivity.

CONCLUDING REMARKS

The study of psychiatric disorders remains one of the great challenges in modern genetic research. Optimism stems from the fact that disease loci have been identified in other complex disorders such as type II diabetes, coronary heart disease and breast cancer, and also from the discovery of highly polymorphic markers throughout the entire genome. This optimism has been added to by work on AD, with the identification of mutations of the APP on chromosome 21, linkage to a second locus on chromosome 14 and the association with apoE E4. Further progress in the functional psychoses may be dependent upon the finding of cytogenetic or molecular neurobiological clues indicating the likely location or identity of disease genes. However, the increase in the number of groups taking a systematic approach to the 'screening' of the genome, chromosome by chromosome, and collaborations in both Europe and North America, may themselves bring about positive findings. With the increase in our understanding of the molecular biology of brain systems, it is likely that a candidate gene approach will become more feasible. For example, the recent cloning of genes for the dopamine receptors D3, D4 and D5 enables the direct testing of genetic hypotheses involving these elements of the dopamine system. Experience in AD shows that it is unwise to exclude candidate loci purely on the basis of negative linkage data where the genetics of a disorder is unclear. The study of families which are individually large enough to give evidence of linkage, and the use of association studies to identify those candidates which contribute only a small amount to the liability to a disease, are both likely to be important.

Success in AD has come initially from restricting analysis to the study of highly familial early-onset pedigrees showing a clear pattern of Mendelian transmission. Subtypes of this sort are not so clear in manic-depression and schizophrenia, and it is important to continue the search for new ways of validating diagnosis and improving the assignment of cases for genetic analysis. On the other hand, final classification of these disorders may well have to wait for the identification of specific gene mutations to clarify the boundaries of the different phenotypes.

REFERENCES

Craddock N and McGuffin P (1993) Approaches to the genetics of affective disorders. *Ann. Med.*, **25**, 317–322.
Crocq MA, Mant R, Asherson P et al. (1992) Association between schizophrenia and homozygosity at the dopamine D3 receptor gene. *J. Med. Genet.*, **29**, 858–860.
Crow TJ (1988) Sex chromosomes and psychosis. The case for a pseudoautosomal locus. *Br. J. Psychiatry*, **153**, 675–683.
Farrer LA, O'Sullivan DM, Cupples LA et al. (1989) Assessment of genetic risk for Alzheimer's disease among first degree relatives. *Ann. Neurol.*, **25**, 485–493.
Goate AM, Chartier-Harlin M-C, Mullan M et al. (1991) Segregation of a missense mutation in the amyloid precursor protein gene with familial Alzheimer's disease. *Nature*, **349**, 704–706.
Gottesman II (1991) *Schizophrenia Genesis: Origins of madness*. San Fransisco: Freeman.
Gottesman II and Bertelsen A (1989) Confirming unexpressed genotypes for schizophrenia. *Arch. Gen. Psychiatr.*, **1**, 287–296.
Mankoo B, Sherrington R, Brynjolfsson J et al. (1991) New microsatellite polymorphisms provide a highly polymorphic map of chromosome 5 bands q11.2–q13.3 for linkage analysis of Icelandic and English families affected by schizophrenia. *2nd World Congress Psychiatr. Genet.*, **2**, 17.
McGuffin P (1991) Genetic models of madness. In McGuffin P and Murray R (eds), *The New Genetics of Mental Illness* (pp. 27–43). Butterworth-Heinemann.
McGuffin P, Sargeant MP, Hett G et al. (1990) Exclusion of a schizophrenia susceptibility gene from the chromosome 5q11–q13 region: new data and a reanalysis of previous reports. *Am. J. Hum. Genet.*, **47**, 524–535.
Owen MJ (1994) The molecular genetics of Alzheimer's disease. In Owen F and Itzhaki R (eds), *Molecular and Cell Biology of Neuropsychiatric Diseases* (pp. 92–109). London: Chapman & Hall.

Owen MJ and McGuffin P (1992) The molecular genetics of schizophrenia: blind alleys, acts of blind faith and difficult science. (Editorial.) *Br. Med. J.*, **305**, 664–665.

Owen MJ and McGuffin P (1993) Association and linkage: complementary strategies for complex disorders. *J. Med. Genet.*, **30**, 638–639.

Prusiner SB (1994) Prion diseases of humans and animals. *J. R. Col. Physicians Lond.* (suppl.), 28.

Saunders AM, Strittmatter WJ and Schmechel D (1993) Association of apolipoprotein E allele 4 with late-onset familial and sporadic Alzheimer's disease. *Neurology*, **43**, 1462–1472.

Sobell JC, Heston LL and Sommer SS (1992) Delineation of genetic predisposition to multifactorial disease: a general approach on the threshold of feasibility. *Genomics*, **12**, 1–6.

St George Hyslop PH, Tanzi RE, Polinsky RJ *et al.* (1987). The genetic defect causing familial Alzheimer's disease maps on chromosome 21. *Science*, **235**, 885–890.

Index

Note—page numbers in *italics* refer to figures and tables

Index compiled by Jill Halliday